THE HIDDEN UNIVERSE WITHIN

How Trillions of Microbes Control Your Mind, Body, and Destiny

By

Mohammad Umar

FIRST EDITION — 2026

PREFACE

A Letter to the Reader

There is a moment in every scientist's life when the ground shifts beneath their feet. Not literally, of course, but intellectually, philosophically, in that deep interior space where assumptions once lived like old furniture, comfortable and unquestioned. For me, that moment came in a laboratory, staring at a microscope slide that revealed something so simple yet so staggering that I had to sit down and simply breathe. What I was looking at was not a disease, not a pathogen, not something dangerous. What I was looking at was life. Extraordinary, ancient, and completely invisible to the naked eye. Billions upon billions of microorganisms, going about their business, completely indifferent to my astonishment. And yet, somehow, they were inside me. Inside all of us. All the time.

This book was born from that astonishment. It was also born from years of frustration with how we talk about microbes in popular culture. We live in a world that has declared war on bacteria. We have antibacterial soaps, antibacterial wipes, antibacterial everything. We spray, scrub, and sanitize as though the greatest threat to human civilization is the invisible world lurking on our doorknobs and kitchen counters. And yet the science, the real, peer-reviewed, rigorously tested science, tells a completely different story. The story is not one of war. It is one of partnership. Of co-evolution. Of a relationship so ancient and so intimate that it has literally shaped who we are as a species.

The human microbiome, the collection of all the microorganisms that live on and inside the human body, is the subject of what may be the most exciting scientific revolution of the twenty-first century. In the last two decades alone, researchers have discovered that the bacteria, viruses, fungi, and other microscopic organisms that inhabit us are not merely passengers. They are pilots. They influence our digestion, our immune system, our metabolism, our hormones, our moods, and possibly even our thoughts and decisions. They communicate with our brains through a network so sophisticated that scientists have given it its own name: the gut-brain axis. They shape our children's development from the moment of birth. They determine, in part, whether we will be obese or lean, anxious or calm, vulnerable to cancer or resistant to it.

This is not science fiction. This is science. And it is happening right now, in laboratories and hospitals and research institutions around the world, quietly rewriting everything we thought we knew about human biology, human health, and human identity.

I wrote this book because I believe that everyone deserves to understand this revolution. Not just scientists and doctors, but teachers and parents and students and curious people who lie awake at night wondering how the universe works. The science of the microbiome is not just a medical story. It is a philosophical story. It asks us to reconsider what it means to be an individual. If trillions of non-human organisms are shaping your mood, your appetite, your immune response, and perhaps even your personality, then what exactly is the self? Who is in charge? The answers, as you will discover, are more fascinating than any science fiction writer could invent.

Each chapter of this book takes a different angle on the microbiome story. Some chapters are written in the style of a scientific adventure, following discoveries as they happened. Others are more meditative, philosophical, asking big questions about identity and existence. Still others are practical guides, written with the urgency of a doctor speaking directly to a patient. This variety is intentional. The microbiome is a subject too large, too rich, and too important to be contained in a single voice or a single style. It deserves to be explored from every direction.

I want to be clear about what this book is not. It is not a diet book. It is not a collection of easy answers or miracle cures. The science of the microbiome is still young, and there is much we do not yet know. Where the research is solid, I say so with confidence. Where it is preliminary or contested, I say that too. I have tried, throughout every page, to be honest about the difference between what we know and what we merely suspect, between the science and the speculation. That distinction matters enormously in a field that has attracted as much hype as it has genuine discovery.

What you hold in your hands is an invitation to see yourself differently. To understand that you are not a single organism moving through the world alone, but an entire ecosystem, a living, breathing community of trillions of beings, human and non-human, working together in a collaboration so ancient it predates the first forests, the first fish, the first flowers. You are, in the most literal sense possible, a universe unto yourself. And that universe is worth understanding.

Welcome to the hidden world within.

Mohammad Umar

INTRODUCTION

The Revolution Beneath the Skin

"To call yourself an individual is to tell only half the story. The other half is written in organisms too small to see, too numerous to count, and too important to ignore."

— Mohammad Umar

I. A Science Born in Astonishment

Every great scientific revolution begins with a moment of profound discomfort. Not the ordinary discomfort of a difficult equation or a recalcitrant experiment, but something deeper: the vertigo of discovering that the world is not organized the way we thought it was. Copernicus caused it when he placed the sun at the center of the solar system. Darwin caused it when he traced the ancestry of human beings back to the same primordial source as every other living thing. Watson and Crick caused it when they revealed that the entirety of biological heredity was written in an elegant double helix of just four chemical letters.

We are living through another such moment now. It is less dramatic in its initial announcement than those earlier revolutions. It did not arrive on a single date with a single paper or a single name attached. It crept up on us gradually, through thousands of studies published in dozens of journals by hundreds of researchers across dozens of countries, accumulating over two decades into a body of evidence so substantial and so consistent that its implications can no longer be avoided or minimized. The revolution is this: the human body is not primarily human.

For every human cell in your body, there is at least one microbial cell: a bacterium, a virus, a fungus, an archaeon, or another microscopic organism that is living in or on you right now, at this very moment, as you read these words. The collective genome of these microorganisms, their total genetic library, contains roughly 150 times more genes than your human genome. The collective metabolic activity of the gut microbiome alone, the community of microorganisms living in your digestive tract, rivals that of the liver in its complexity and importance. The microbes that inhabit you are not parasites, not pathogens, not incidental contaminants from an unclean world. They are partners. They are, in the most biologically rigorous sense of the word, part of you.

This book is about that partnership: where it came from, how it works, what happens when it breaks down, and what it means for the future of human health. It is also, inevitably, about what this partnership means for our understanding of ourselves. Because once you truly grasp the biological reality of the microbiome, the old story of the individual, the self-contained human being with a clear boundary between self and other, no longer holds. A new story is needed. And the new story is, I believe, far more interesting than the old one.

II. Why This Moment, Why This Book

The science of the human microbiome is not entirely new. Antonie van Leeuwenhoek observed bacteria in dental scrapings in the 1670s. Louis Pasteur identified microorganisms as agents of fermentation and disease in the nineteenth century. Elie Metchnikoff, the Russian zoologist who shared the 1908 Nobel Prize in Physiology or Medicine, speculated that the bacteria of the gut might have a fundamental influence on human health and longevity, and recommended the regular consumption of fermented milk as a way of cultivating a beneficial microbial community. He was laughed at by many of his contemporaries. He was, as we now know, essentially right.

What has changed in the last two decades is not the basic idea that microbes matter to human health. It is our ability to see them clearly, to count them, to identify them, to map their functions, and to understand the molecular mechanisms through which they interact with human biology. The key technological development was the introduction of next-generation DNA sequencing techniques in the mid-2000s, which made it possible to sequence the entire genetic contents of a microbial community without the need to culture individual organisms in a laboratory. This was a crucial breakthrough, because the vast majority of gut bacteria cannot be cultured in the laboratory under standard conditions. Before sequencing-based approaches, we were effectively blind to most of the microbial life inside us. The sequencing revolution gave us our first clear view.

The Human Microbiome Project, launched by the United States National Institutes of Health in 2007, put the full resources of modern genomics to work on the problem. Over nearly a decade, it produced a comprehensive map of the microbial communities inhabiting healthy human bodies and laid the empirical foundation for everything that has come since. What has come since is extraordinary: the discovery of the gut-brain axis, the revelation that the microbiome influences mood, anxiety, and possibly the development of conditions from depression to autism; the demonstration that gut bacteria play a central role in obesity and metabolic disease; the finding that cancer immunotherapy outcomes are powerfully shaped by the patient's gut microbiome; the development of fecal microbiota transplantation as a therapy for conditions ranging from Clostridioides difficile infection to inflammatory bowel disease. Each year brings new discoveries that further expand the known territory of the microbiome's influence on human health.

This is the right moment for a book that brings all of this science together in a form that is accessible to anyone who wants to understand it. Not a specialist treatise, not a simplified pop-science overview that strips out the complexity and nuance, but a serious, deeply informed, narrative exploration of what we now know about the microbiome and what it means for how we understand our bodies, our health, and our place in the living world. That is what this book attempts to be.

III. What You Will Find in These Pages

This book is organized into five parts, each exploring a different dimension of the microbiome story. Part One, The Invisible World, establishes the foundations: the deep evolutionary history of microbial life, the remarkable complexity of the human microbiome, and the fascinating story of how each of us acquires our unique microbial community, beginning at birth and continuing throughout our lives.

Part Two, The Microbiome and the Body, examines the central roles the microbiome plays in human physiology: its command center function in the gut, its extraordinary partnership with the immune system, and its profound and still-surprising influence on metabolism and body weight. These chapters reveal the microbiome not as an accessory to human biology but as one of its primary engines.

Part Three, The Gut-Brain Axis, takes us into what is perhaps the most astonishing territory in all of microbiome science: the communication network linking the gut and the brain. You will encounter the enteric nervous system, the brain inside your belly; discover that 95 percent of the body's serotonin is made in the gut, not in the brain; and follow the emerging evidence linking the gut microbiome to depression, anxiety, stress responses, sleep, and some of the most complex neurological and psychiatric conditions in medicine.

Part Four, The Microbiome and Disease, confronts the dark side: what happens when the inner ecosystem breaks down. It covers the microbiome's connections to cancer, including the extraordinary finding that immunotherapy outcomes depend on the gut microbiome, and the deeply troubling legacy of antibiotic overuse, which has simultaneously saved hundreds of millions of lives and created a global resistance crisis while degrading the microbiomes of entire populations.

Part Five, The Future of the Microbiome, is the most forward-looking section of the book. It covers fecal microbiota transplantation, one of the most effective medical procedures ever developed despite its challenging nature; separates the science from the marketing in the world of probiotics and prebiotics; explores the coming era of personalized microbiome medicine; and concludes with a practical, evidence-based guide to protecting and restoring your own inner ecosystem.

Each chapter is written in a different voice and style, a deliberate choice that reflects the multidimensional nature of the subject. The microbiome story is too large and too rich to be told in a single key. Some chapters are sweeping and historical, others are intimate and clinical, others are philosophical and challenging. Some begin with scientific experiments; others begin with the personal experiences of patients or researchers. What unites them is the underlying commitment to honesty about what we know, what we suspect, and what remains genuinely unknown.

IV. On the Science and Its Limits

A word of honesty about the science itself, because it matters enormously in a field that has attracted as much enthusiasm, speculation, and commercial exploitation as the microbiome has.

Microbiome science is real, and it is important. The findings described in this book are drawn from peer-reviewed research published in some of the world's most rigorous scientific journals. The mechanisms described are, in many cases, well-established and backed by multiple independent lines of evidence. The clinical applications are, in some cases, already in use in hospitals and clinics around the world. None of this is pseudoscience or wishful thinking.

But microbiome science is also young, and youth in science means uncertainty. Many of the most exciting findings are correlational rather than causal: we have observed that people with condition X have microbiome Y, but have not always proven that microbiome Y causes condition X rather than resulting from it or simply accompanying it. Animal studies, particularly in mice, have provided crucial mechanistic insights, but the translation from mouse biology to human biology is always uncertain and sometimes fails entirely. Human clinical trials of microbiome-based interventions are still in relatively early stages for most conditions outside of Clostridioides difficile infection. The field is moving rapidly, which means that some of what is considered cutting-edge today may be overturned, refined, or significantly complicated by tomorrow's research.

Throughout this book, I have tried to be explicit about the difference between what is firmly established, what is strongly suggested by available evidence, and what is still preliminary or speculative. I have tried to distinguish between findings in animal models and findings in humans, and to note where human evidence is robust versus where it is limited. I have tried not to oversell the science, because overselling science, particularly in the domain of health and medicine, does real harm: it creates unrealistic expectations, fuels a marketplace of dubious products and interventions, and ultimately undermines trust in legitimate research when the overpromised outcomes fail to materialize.

The microbiome does not need to be oversold. Its genuine importance, already established by rigorous research, is remarkable enough. This book tries to convey that importance faithfully, with both appropriate excitement and appropriate epistemic humility.

V. A Personal Note

I came to the microbiome through an unexpected door. I was not trained as a microbiologist. My background is in science writing and the history of biology, and I came to this subject the way most science writers come to their best subjects: through a combination of professional assignment and personal obsession. The more I read, the more I found that the microbiome was not merely a scientific topic but a genuinely philosophical one, a set of discoveries that forced me to reconsider fundamental assumptions about what I am, where I begin and end, and how my choices ripple outward into the biological community I carry with me everywhere I go.

That philosophical dimension is what I have tried to bring to this book alongside the science. The question of what the microbiome means for human identity and agency is, I believe, as important as the question of what it means for human health. If the microbes inside us shape our moods, our cravings, our stress responses, and possibly our decisions, then the old picture of the autonomous individual, making rational choices based on a unified and internally consistent self, becomes significantly more complicated. We are not diminished by this complexity. We are enlarged by it. The recognition that we are ecosystems, not islands, is not a loss of selfhood but an expansion of the story we tell about what it means to be a self.

I hope this book leaves you with that sense of expansion. I hope it leaves you not with a list of things to eat or avoid, though it will certainly give you some of those, but with a fundamentally different way of seeing your own body and your relationship to the living world around you. The microbes were here before us. They built the world we live in. They made possible the oxygen we breathe, the soil that grows our food, the ecosystems that sustain all complex life on Earth. And they live inside us, in their billions, doing work we are only beginning to comprehend. They deserve our understanding. And in understanding them, we come, at last, to understand ourselves a little better.

VI. How to Read This Book

This book is designed to be read from beginning to end, as each part builds on the foundations laid by the one before it. However, for readers who have specific interests, it can also be approached thematically. Those primarily interested in mental health and the gut-brain connection may wish to go directly to Part Three after reading the foundational Part One. Those most interested in practical guidance for improving their gut health will find Part Five most immediately useful, though it will be richer and more comprehensible with the context provided by the earlier parts. Those primarily interested in the future of medicine and the clinical applications of microbiome science may wish to focus on Parts Four and Five.

I have tried to write for a general reader with no scientific background required, explaining technical concepts clearly when they arise and never assuming prior knowledge of biology, chemistry, or medicine. At the same time, I have not simplified the science to the point of inaccuracy. Where the real biology is genuinely complex, I have honored that complexity rather than hiding it behind reassuring oversimplification. Science is not always tidy or intuitive, and the microbiome is no exception. The effort required to understand it accurately is, I promise, worth it.

The glossary at the back of the book provides clear definitions of all technical terms used in the text, and the appendices include guidance on further reading, the key scientists and institutions driving this research, and a practical 30-day framework for those who wish to begin working with the science immediately in their own lives.

One final note: the science in this book is drawn from published research, and the researchers, studies, and institutions that I have cited or paraphrased are, to the best of my knowledge, accurately represented. Any errors of fact or interpretation are my own, and I welcome correction from readers who are specialists in relevant fields.

VII. The Stakes

Why does any of this matter? Why should the composition of the bacteria in your colon be of concern to anyone beyond specialists in gastroenterology?

It matters because we are facing a global health crisis whose roots, the science now strongly suggests, are at least partly microbial. The epidemic of chronic inflammatory disease that characterizes the industrialized world in the twenty-first century, the soaring rates of obesity, type 2 diabetes, inflammatory bowel disease, allergies, asthma, autoimmune conditions, and mental health disorders, is coinciding with and is at least partially driven by a global degradation of the human microbiome. The forces driving that degradation are the same forces that define modern life: industrial food production, antibiotic overuse, sanitized indoor environments, separation from natural green spaces, chronic stress, and disrupted sleep. We have inadvertently been conducting a large-scale experiment on our inner ecosystems, and the results are coming in, and they are not encouraging.

Understanding this is the first step toward addressing it. Not with a single miracle intervention, not with a probiotic pill or a fecal transplant, though those tools have their place, but with a comprehensive rethinking of how we live, how we eat, how we design our environments, and how we use the medical technologies that have given us such extraordinary power over infectious disease while creating new vulnerabilities we are only beginning to appreciate.

The stakes of getting this right are large. The global burden of chronic inflammatory disease is already catastrophic in human terms and economically ruinous. A future in which microbiome science guides more intelligent approaches to diet, medicine, environmental design, and the management of the microbial world could represent one of the most significant improvements in human health since the antibiotic revolution itself. A future in which we continue to ignore or misunderstand the microbiome risks compounding the chronic disease epidemic with new and unanticipated microbial crises.

We are, as a civilization, at an inflection point in our relationship with the microbial world. We can choose to engage with that world more intelligently, more humbly, and more collaboratively than we have in the past century of antibiotic medicine and industrial food production. Or we can continue to treat the microbial world as an enemy to be defeated, ignoring the evidence that it is, far more fundamentally, a partner to be understood and respected.

The choice, as this book will make clear, is not merely a scientific or medical one. It is a civilizational one. And it begins, as so many important things begin, with understanding.

Mohammad Umar

PART ONE

THE INVISIBLE WORLD

Discovering the Universe We Never Knew We Carried

CHAPTER ONE

The World Before Us

The Age of Microbes and the Planet They Built

Imagine standing at the edge of a vast ocean. Not the ocean as we know it today, with its coral reefs and whales and shimmering schools of fish. Imagine an ocean from 3.8 billion years ago: still, darkly warm, its surface struck by frequent lightning, its depths filled with chemicals and gases and heat rising from underwater volcanic vents. The sky above is orange and toxic. There is no oxygen in the air. No plants on the land, because there is no land worth speaking of, just bare rock and shallow, turbid pools. The Earth is young and raw and utterly inhospitable to anything we would recognize as life. And yet, something remarkable is about to happen. Something that will change this planet so completely that billions of years later, creatures of unimaginable complexity will walk upon its surface, breathe its air, drink its water, and call it home.

That something is a single cell. Tiny beyond imagination. Simpler in structure than the most primitive thing we can see with the naked eye. But alive. Unmistakably, miraculously alive. This is the ancestor of every living thing on Earth. This is where our story begins.

The First Inhabitants of Earth

The origin of life is still one of science's deepest mysteries. We do not know with certainty how that first cell came to exist. The leading hypotheses involve complex chemistry occurring in hydrothermal vents on the ocean floor, where superheated water rich in minerals and organic compounds bubbled up from the Earth's interior. In these environments, the raw chemical precursors of life, amino acids, nucleotides, lipids, may have spontaneously assembled into the first self-replicating molecules, the earliest ancestors of DNA and RNA. From there, the journey to the first living cell was still enormously long and uncertain, involving countless chemical steps we can only partially reconstruct. But at some point, perhaps around 3.8 billion years ago, or possibly even earlier, the first true cellular life appeared on Earth.

These earliest life forms were prokaryotes, cells without a defined nucleus, far simpler in structure than any cell in your body today. They had no mitochondria, no complex internal machinery. They were essentially just a membrane enclosing a collection of molecules capable of capturing energy, growing, and copying themselves. By the standards of modern biology, they were rudimentary to the point of absurdity. And yet they were profoundly successful. They filled the oceans. They colonized the rocky shores. They evolved at dizzying speed, driven by mutation and natural selection, diversifying into thousands, then millions of distinct forms, each adapted to exploit a different chemical niche in the environment.

For approximately two billion years, these microorganisms were the only life on Earth. Two billion years. To put that number in perspective: that is roughly half the entire age of our planet. During that inconceivably long span of time, microbes were alone, the undisputed masters of the Earth, and they were doing something that would ultimately make every other form of life possible. They were engineering the planet.

The Engineers of Our World

Perhaps the most astonishing thing about early microbial life is what it did to the Earth's atmosphere. The early atmosphere of our planet contained no free oxygen. Life as we know it today, from butterflies to blue whales to human beings, is fundamentally dependent on oxygen for its metabolism. Without it, we would be dead in minutes. And yet the Earth existed for over a billion years with essentially no atmospheric oxygen at all. The transformation came from microbes.

Around 2.7 billion years ago, a group of bacteria called cyanobacteria evolved the ability to perform an entirely new kind of metabolism: photosynthesis. Using sunlight, water, and carbon dioxide, they began producing organic molecules to fuel their growth. And as a byproduct of this process, they released oxygen into the water and then into the atmosphere. At first, this oxygen was quickly absorbed by dissolved iron in the oceans, forming the banded iron formations we find in ancient rocks today. But eventually, the cyanobacteria produced so much oxygen that it began to accumulate in the atmosphere itself. This process, known as the Great Oxidation Event, occurred around 2.4 billion years ago and stands as perhaps the most radical environmental transformation in Earth's history. What was once a poison to the anaerobic organisms that had dominated life for over a billion years became the foundation of an entirely new kind of biology, oxygen-fueled life that was more energetically powerful than anything that had come before.

Microbes did not stop at reshaping the atmosphere. They also played a foundational role in creating the soil that covers the Earth's continents. The breakdown of bare rock into the rich, nutrient-laden soil that supports plant life on land is in part a microbial achievement. Bacteria and fungi colonized rock surfaces, secreting acids that dissolved minerals, breaking down the stone into particles that gradually accumulated into soil. They fixed nitrogen from the atmosphere into forms that plants could use, making plant life on land not just possible but spectacularly productive. In the oceans, marine microbes form the base of virtually every food chain, forming the phytoplankton blooms that feed the zooplankton, which feed the fish, which feed everything above them. Remove the microbes from the ocean and the entire ocean food web collapses.

This is the perspective that science now asks us to hold about the microbial world: not as our enemy, not as the cause of disease and death, though some microbes certainly are that. But as the architects of the living planet. Without microbes, there is no oxygen in the air, no soil on the land, no food in the ocean, no forest, no meadow, no reef. The planet we know is a microbial creation.

The Discovery That Changed Everything

For most of human history, the microbial world was completely invisible to us. Not because people did not wonder about invisible causes of disease or fermentation or decay, they did, but because they lacked the tools to see. That changed in the 1670s, in the Dutch city of Delft, where a draper named Antonie van Leeuwenhoek ground lenses of extraordinary quality and began peering through them at the world around him. What he saw astonished him beyond expression.

In a droplet of pond water, he observed tiny creatures moving with apparent purposefulness. In the scrapings from his own teeth, he found a menagerie of living things so small that thousands could fit on the head of a pin. He called them animalcules, little animals, and he wrote detailed descriptions of them to the Royal Society of London with a mixture of excitement and barely contained disbelief. The scientists of Europe received his reports with a mixture of fascination and deep skepticism. The idea that an entire invisible world of living creatures existed everywhere around us, on us, inside us, was simply too strange to accept without hesitation.

Leeuwenhoek was right, of course. Over the centuries that followed, as microscopy improved and then transformed into entirely new technologies, scientists gradually revealed the extraordinary diversity and ubiquity of the microbial world. They discovered that bacteria are not simply primitive, undifferentiated blobs of life but organisms of remarkable complexity, capable of communication, cooperation, and adaptation that continues to astonish researchers today. They discovered viruses, particles so stripped-down that scientists still debate whether they are truly alive, and yet so influential that they have shaped the evolution of every complex organism on Earth. They discovered fungi and archaea and protists and a whole catalog of life forms that exist beneath the threshold of visibility but above the threshold of biological significance.

Masters of Survival

One of the most remarkable qualities of microbial life is its extraordinary resilience. Microbes can survive in conditions that would instantly kill any plant or animal. There are bacteria living in the superheated water of hydrothermal vents at the bottom of the ocean, where temperatures exceed 120 degrees Celsius, hot enough to boil water at normal pressure. There are bacteria living in the cold, dry valleys of Antarctica, where temperatures drop to minus 60 degrees and liquid water is almost entirely absent. There are bacteria in the heart of nuclear reactors, thriving in radiation levels that would shred human DNA in seconds. There are bacteria living miles underground in solid rock, metabolizing on geological timescales, essentially unchanging for millions of years. There are bacteria in the upper atmosphere, carried on wind currents high above the clouds.

This resilience is not accidental. It is the product of 3.8 billion years of evolution, of an unimaginable number of generations and mutations and selections, refining and diversifying the toolkit of microbial survival until it could cope with almost any environment Earth could offer. It is this same evolutionary deep-rootedness that makes the microbial world so important to understand. When we speak of the human microbiome, we are not speaking of something peripheral or incidental to human biology. We are speaking of the culmination of billions of years of co-evolution, of an ancient partnership between microbial and animal life that has been shaping organisms like us for hundreds of millions of years.

The Scale of the Invisible

Numbers, when they get large enough, cease to feel real. We say that the universe contains two trillion galaxies and the mind simply nods and moves on, unable to truly grasp what two trillion means. The numbers involved in the microbial world are similarly staggering and similarly difficult to hold in the imagination. There are an estimated ten to the power of thirty bacteria on Earth. That is a one followed by thirty zeros. If you lined up all the bacteria on Earth end to end, the resulting chain would stretch across the observable universe and back multiple times. The combined mass of all bacteria on Earth significantly exceeds the combined mass of all animals, all plants, and all fungi combined. Bacteria are not a footnote in the story of life on Earth. They are the central text.

And they are everywhere. Every surface you touch is covered in a film of microbial life. Every breath you take brings bacteria and viruses into your lungs. Every meal you eat carries microorganisms into your digestive system. The air itself is a medium through which microbial life travels constantly. There is no sterile environment in nature, no corner of the living world from which microbes are absent. We live inside the microbial world as fish live in the ocean. It surrounds us, permeates us, sustains us, and has done so for the entire history of our species.

This is the world we are entering as we begin our exploration of the human microbiome. Not a world of danger and disease, though we will not shy away from those aspects of the story. But a world of extraordinary complexity and beauty, of ancient relationships and modern revelations, of science pushing at the boundaries of what we thought we knew about life, about health, and about ourselves. The microbes were here first. In many ways, they are still in charge. And understanding them, truly understanding them, may be the most important thing we can do for the future of human health.

CHAPTER TWO

You Are Not Alone

Meet Your Microbiome

You have been lied to, not maliciously, not with intent to deceive, but by the quietly accumulated assumptions of a culture that has spent the better part of a century thinking about the human body in the wrong way. The lie goes something like this: you are a single organism. One individual. A coherent, unified being with a clear boundary between yourself and the world. Your skin is a wall. Everything inside it is you. Everything outside it is not you. Clean, simple, and deeply, profoundly, scientifically incorrect.

Here is the truth. Right now, as you read these words, there are approximately 38 trillion microbial cells living on and inside your body. Bacteria, viruses, fungi, archaea, and other microorganisms are inhabiting your gut, your skin, your mouth, your lungs, your reproductive tract, and many other tissues that scientists are only beginning to explore. These organisms are not hitchhikers or stowaways. They are partners, and in many cases, they are doing jobs that your own human cells cannot do. You are, in the most biologically accurate sense of the word, not an individual. You are an ecosystem.

Defining the Microbiome

The word microbiome was coined in the early 2000s and has since become one of the most talked-about terms in all of medicine. But what exactly does it mean? The term refers to the collection of all microorganisms that inhabit a particular environment. In the context of human biology, the human microbiome refers to all the bacteria, viruses, fungi, archaea, and other microscopic life forms that live in and on the human body, along with their collective genetic material. That last part is important, because the microbiome is not merely a collection of organisms. It is a collection of genes. And the number of microbial genes in the human body is extraordinary.

The human genome, the full set of genetic instructions encoded in human DNA, contains roughly 20,000 protein-coding genes. By most estimates, the collective genome of the human microbiome, sometimes called the metagenome or the second genome, contains approximately 3 million genes. That means the microbiome contributes roughly 150 times more genetic information to the human body than the human genome itself does. In a very real sense, the human body is not primarily human in its genetic makeup. It is primarily microbial. The implications of this are still being worked out by scientists around the world, but they are almost certainly profound.

The Human Microbiome Project

The scientific effort to understand the human microbiome began in earnest with a landmark initiative launched by the United States National Institutes of Health in 2007: the Human Microbiome Project. It was in many ways the microbial equivalent of the Human Genome Project, the massive international collaboration that had completed the first full mapping of human DNA in 2003. Where the Human Genome Project asked what genes the human body contained, the Human Microbiome Project asked who else was living there.

Over the course of nearly a decade, researchers at dozens of institutions across the United States collected samples from 300 healthy adult volunteers, swabbing 18 different body sites including the gut, mouth, skin, and urogenital tract, and then using advanced genetic sequencing technologies to identify the microbial communities living at each location. The results were revelatory. Scientists discovered that the human microbiome was far more diverse, far more complex, and far more variable from person to person than anyone had anticipated. They found that healthy people could have wildly different microbial communities and still be equally healthy. They discovered that certain body sites, like the gut, were dominated by consistent bacterial phyla, while others, like the skin, showed enormous variation depending on location. Most importantly, they established the baseline: what a healthy human microbiome looks like. Without that baseline, it would be nearly impossible to understand what a diseased one looks like.

A Tour of Your Microbial Residents

The microbiome is not a single, uniform community. It is a collection of distinct ecosystems, each tailored to the specific conditions of its habitat. The gut microbiome is the largest and most studied. Your digestive tract, particularly the large intestine, contains somewhere between 500 and 1,000 distinct bacterial species, living in a community of staggering density. Some estimates suggest that there are more bacteria in a single gram of human feces than there are stars in the Milky Way galaxy. The dominant players are two major groups: the Firmicutes and the Bacteroidetes, which together account for more than 90 percent of the bacteria in the average adult gut. Within those broad groups, thousands of species compete, cooperate, and carve out their own ecological niches, creating a community of almost unimaginable complexity.

The skin microbiome is the second largest and, in some ways, the most intimate. Your skin is covered in microbial life, approximately 1.5 trillion microbes, though the number and composition vary enormously depending on the body site. The moist folds of the elbow harbor very different communities than the dry, exposed surface of the forearm. The oily environment of the forehead supports different species than the cooler, drier landscape of the lower leg. Far from being a problem, this skin microbiome plays a crucial role in defense: the bacteria that live on your skin actively compete with dangerous pathogens for space and resources, and they produce compounds that inhibit the growth of harmful organisms. When the skin microbiome is disrupted, conditions like eczema, acne, and psoriasis often follow.

The oral microbiome is perhaps the most socially consequential. Your mouth contains approximately 700 different species of bacteria, making it one of the most diverse microbial habitats in the entire body. These oral bacteria are responsible for both the good and the bad of oral health. Some of them protect against disease, while others, particularly in the presence of sugar and in the absence of adequate oral hygiene, produce acids that erode tooth enamel and cause cavities. But the oral microbiome's significance extends far beyond dental health. Research has linked oral bacteria to cardiovascular disease, diabetes, and pregnancy complications. Bacteria from the mouth can enter the bloodstream and travel to distant organs. The state of your mouth, in a very real sense, reflects and influences the state of your entire body.

Even the lungs, once thought to be sterile, have their own microbiome. For decades, medical textbooks taught that healthy lungs contained no bacteria. We now know this is incorrect. The lungs harbor a small but distinct community of microorganisms, and disruptions to this community are associated with respiratory diseases including asthma, chronic obstructive pulmonary disease, and possibly even lung cancer. The vaginal microbiome, present in approximately half the global population, is one of the least diverse but most important microbial communities in the body. Dominated by a few species of Lactobacillus bacteria, a healthy vaginal microbiome maintains an acidic environment that protects against sexually transmitted infections and is crucial for healthy pregnancy outcomes.

Your Microbiome Is Uniquely Yours

One of the most striking findings of microbiome research is just how individual the microbiome is. Like a fingerprint, your microbial community is unlike anyone else's. Even identical twins, who share 100 percent of their human DNA, have significantly different microbiomes. This individuality arises from the unique combination of factors that have shaped your microbial community across your lifetime: the circumstances of your birth, the way you were fed as an infant, the environments you were exposed to as a child, the foods you have eaten, the medications you have taken, the illnesses you have suffered, the people and animals you have lived with, and countless other influences, each leaving its mark on the composition of your inner microbial world.

This individuality has profound implications. It means that two people can eat the same meal and have completely different metabolic responses to it, because their gut bacteria process the food differently. It means that a drug that works brilliantly for one patient may fail entirely in another, partly because of differences in their microbiomes that alter how the drug is metabolized and what effects it has on the body. It means that the search for universal prescriptions for gut health, the idea that everyone should eat the same things, take the same probiotics, follow the same protocols, is fundamentally misguided. The future of microbiome medicine will be personalized, tailored to the unique microbial landscape of each individual.

What Your Microbiome Does for You

The functions of the microbiome are so numerous and so central to normal biological function that it is almost easier to ask what the microbiome does not do than what it does. At the most basic level, gut bacteria help digest food. Many of the complex carbohydrates we eat, the fibers from vegetables, fruits, and whole grains, cannot be broken down by human enzymes. Without gut bacteria to ferment these fibers, they would pass through the digestive system largely unused. The bacteria break them down into short-chain fatty acids, particularly butyrate, acetate, and propionate, which are then absorbed into the bloodstream and used as energy by the body's cells. Butyrate, in particular, is the primary energy source for the cells lining the colon and plays a critical role in maintaining the integrity of the gut barrier.

Gut bacteria also synthesize vitamins that the human body cannot produce on its own. Vitamin K, essential for blood clotting, is produced in significant amounts by gut bacteria. Several B vitamins, including B12, folate, and biotin, are also produced by intestinal microbes. These microbial contributions to human nutrition are not trivial supplements; they are essential inputs without which normal human metabolism would fail. The microbiome is, in a very real sense, a metabolic organ, one that works alongside the liver, the kidneys, and the other organs of the body to keep human biology running.

Perhaps most importantly, the microbiome educates and regulates the immune system. Roughly 70 percent of the body's immune cells are located in and around the gut, and the microbial community there is in constant conversation with them. From the earliest moments of life, gut bacteria send chemical signals to immune cells that teach them how to distinguish harmless proteins from dangerous pathogens, how to mount an appropriate inflammatory response without causing collateral damage, and how to stand down when the threat has passed. Without this microbial education, the immune system develops in a disordered, hyperreactive way that leads to allergies, autoimmune disease, and chronic inflammation. The microbiome, in this sense, is not just a tenant in the human body. It is a teacher, a regulator, a co-pilot in the management of human health.

CHAPTER THREE

How You Got Your Microbiome

The Story of Your Microbial Birth

"We are born into microbes, nurtured by microbes, and shaped by microbes from our very first breath. Our microbial story begins not at birth, but perhaps even before it."

— Dr. Maria Gloria Dominguez-Bello, Microbiologist

Every human being carries a story. Not just the story of their parents, their culture, their language, and their experiences, but a deeper, more ancient story written in the language of microbial ecology. It is the story of how you were colonized, how your inner universe was first populated, shaped, and diversified by the microorganisms that would go on to become some of the most important biological partners of your life. This story begins in a place we rarely consider as a site of biological drama: the birth canal.

Before You Were Born

For decades, the medical community operated under a comfortable assumption: the womb was sterile. A developing fetus, it was believed, existed in a pristine, microbe-free environment until the moment of birth. This belief was so deeply entrenched that it was rarely questioned. It shaped obstetric practice, informed neonatal care, and went largely unchallenged until the tools of modern genomic analysis made it possible to look more closely at what was actually happening inside the uterus.

What researchers found when they did look was surprising. Studies examining the placenta, the amniotic fluid, and even meconium, the first stool passed by a newborn, have detected the genetic signatures of bacteria. This has sparked one of the most vigorous debates in contemporary microbiome science. Some researchers argue that these findings represent genuine evidence of a fetal microbiome, a small community of microorganisms living in the uterus during pregnancy. Others contend that the detected signals are simply contamination, the result of the extraordinary sensitivity of modern sequencing techniques picking up trace amounts of microbial DNA from the environment or from sampling equipment. The debate is ongoing, and the science is not yet settled.

What is much clearer is what happens during and after birth. However quiet the prenatal microbial story might be, the moment of birth is a microbial explosion, a dramatic seeding event in which a vast community of microorganisms from the mother colonizes the newborn child in the span of hours. And the circumstances of that birth, whether it is vaginal or by cesarean section, whether the infant is immediately placed skin-to-skin with the mother, whether it is breastfed or formula-fed, have consequences that science is only beginning to fully comprehend.

The Birth Canal: A River of Life

The journey through the birth canal is, among other things, a microbial baptism. As a baby passes through the vaginal canal during a natural birth, it is coated in the mother's vaginal and fecal microbiota. This coating includes a rich community of Lactobacillus bacteria, the same organisms that give yogurt its tangy flavor, along with a diverse array of other bacterial species. The baby swallows these microbes. They enter the nose and lungs with the first breath. They colonize the skin. Within hours of birth, a naturally delivered infant is already home to a complex microbial community that is strongly shaped by the mother's own microbiome.

This is not a design flaw. It is a design. The transfer of the mother's microbiome to the infant is a biological gift, refined over millions of years of evolution, that provides the newborn with a starter community of microorganisms specifically suited to colonizing the infant gut and beginning the long process of immune system education. The Lactobacillus bacteria transferred during vaginal birth are particularly important, because they produce lactic acid, which creates an acidic gut environment that inhibits the growth of many dangerous pathogens. They also produce bacteriocins, antimicrobial compounds that further protect the vulnerable newborn from infection.

The picture changes significantly when a baby is delivered by cesarean section. In a C-section birth, the infant bypasses the birth canal entirely and is born into a very different microbial world. Instead of being coated in the mother's vaginal bacteria, the newborn's first microbial contacts are the bacteria found on the mother's skin, on the surgeon's gloves, in the operating room air, and on the surfaces of the hospital environment. These are very different organisms from those found in the birth canal, and research has consistently shown that C-section babies have significantly different microbiomes from vaginally delivered babies in the days, weeks, and months after birth.

The implications of this difference are a matter of active scientific investigation and considerable debate. Numerous epidemiological studies have found associations between cesarean delivery and higher rates of various health conditions later in life, including asthma, allergies, obesity, and inflammatory bowel disease. The hypothesis is that the altered microbiome resulting from C-section birth may leave gaps in immune system education that increase the risk of these conditions. It is important to emphasize that cesarean birth is sometimes medically necessary and saves lives. The goal of this science is not to stigmatize C-sections but to understand their biological consequences and find ways to mitigate them, such as the practice of vaginal seeding, in which fluids from the mother's vaginal canal are swabbed onto the newborn's mouth and skin immediately after a C-section delivery.

The Gift of Breast Milk

If the birth canal is the first chapter in the story of microbial colonization, breast milk is the second, and it is every bit as rich and complex. Human breast milk was long thought to be essentially sterile, like the womb. Scientists knew it contained antibodies and nutrients, but the idea that it also contained living bacteria was not seriously entertained until the development of modern sequencing technologies made it possible to detect those organisms.

Today, we know that human breast milk contains a remarkably diverse community of bacteria, with estimates ranging from 700 to over 1,000 different species having been detected across different studies. Where do these bacteria come from? Research suggests that at least some of them originate in the mother's own gut and travel via the bloodstream and lymphatic system to the mammary glands, a pathway called the entero-mammary pathway. This is a stunning piece of biological engineering: the mother's gut bacteria are being specifically recruited and transported to the breast in order to be delivered to the nursing infant.

But the microbial gift of breast milk does not stop with bacteria. Human milk also contains human milk oligosaccharides, complex sugars that are the third most abundant solid component of breast milk after fat and lactose. These sugars cannot be digested by the infant. Human infants lack the enzymes to break them down. So why does breast milk contain so many of them? Because they are not food for the baby. They are food for the baby's bacteria, specifically for species of Bifidobacterium that are among the most important colonizers of the infant gut. These oligosaccharides act as a precisely targeted prebiotic, selectively nourishing the bacterial species that the infant most needs to establish a healthy gut microbiome. It is, in the language of ecology, a remarkable example of coevolution: the human body has evolved to produce food specifically designed to feed the microbes it needs.

Childhood: Building the Microbial Self

In the months and years after birth, the microbiome is a work in progress. It shifts constantly, responding to new foods, new environments, new microbial encounters. The introduction of solid foods is a major turning point. As the infant's diet expands beyond breast milk, the gut microbiome begins to diversify dramatically, colonized by new species able to digest the complex carbohydrates found in plant foods. By the age of two or three, most children have developed a gut microbiome that already shows the hallmarks of an adult community, though full maturation takes several more years.

During this period, the microbial encounters of childhood are crucial. When a toddler puts a muddy hand in their mouth, when they crawl across a floor that has been walked on by family members and pets, when they play in dirt or grass or sand, they are not just being messy. They are acquiring microbes. And those microbes, the environmental bacteria and fungi that colonize the child's skin and gut during early childhood, are critically important for the development of a robust, well-calibrated immune system. The 'old friends' hypothesis,' developed by immunologist Graham Rook, proposes that the human immune system evolved in constant contact with a rich community of environmental microbes, and that it requires continued exposure to these organisms in order to develop and function normally. The dramatic rise in allergies, asthma, and autoimmune diseases in industrialized nations over the past century is, at least in part, the consequence of depriving children of these ancient microbial encounters.

Pets play an important role in this story. Children raised in households with dogs are consistently found to have more diverse gut and skin microbiomes than children raised without dogs. Dogs bring in environmental bacteria from the outdoors, they lick faces, they shed fur containing microbial communities, and they fundamentally alter the microbial environment of any home they inhabit. Studies have found that dog ownership is associated with a significantly reduced risk of childhood asthma, and researchers believe the microbiome effects of dog exposure are a major part of the explanation.

A Microbiome for Life

The microbiome continues to evolve throughout the entire human lifespan. Puberty brings hormonal changes that alter the composition of the skin and gut microbiomes. Pregnancy dramatically reshapes the vaginal and gut microbiomes, with the gut microbiome of the third trimester looking markedly different from that of the first, in ways that may help prepare the body for the metabolic demands of supporting a growing fetus. Aging is associated with a gradual decline in microbiome diversity, a loss of certain beneficial bacterial species, and an increase in others associated with inflammation, a process sometimes called inflammaging.

Diet is the single most powerful ongoing influence on the microbiome throughout adulthood. Studies have shown that the composition of the gut microbiome can change significantly within 24 to 48 hours of a major dietary shift. A person who adopts a plant-rich diet quickly sees an increase in the abundance of fiber-fermenting bacteria. A person who shifts to a diet high in animal protein and fat sees a corresponding decrease in those species and an increase in organisms better adapted to metabolizing fat and protein. The microbiome is, in this sense, always in motion, always responding, always adapting to the conditions of its environment.

This dynamism is both a vulnerability and an opportunity. The vulnerability lies in the fact that bad choices, whether dietary, pharmacological, or lifestyle-related, can quickly degrade the microbiome. The opportunity lies in the fact that good choices can just as quickly restore it. Your microbiome is not your destiny. It is your responsibility. And understanding where it came from is the first step toward deciding where you want it to go.

PART TWO

THE MICROBIOME AND THE BODY

How Microbes Run Your Biology from the Inside Out

CHAPTER FOUR

The Gut

The Command Center of Your Inner Universe

In the architecture of the human body, the gut is often treated as a humble plumbing system, a long tube through which food travels, nutrients are extracted, and waste is expelled. This view is so profoundly inadequate that it is difficult to know where to begin correcting it. The gut is not merely a tube. It is the body's most complex organ, home to its largest immune tissue, the site of its most intensive microbial activity, a producer of dozens of hormones and neurotransmitters, and the location of a second nervous system so elaborate that it can function entirely independently of the brain. To understand human health without understanding the gut is to try to understand a city without understanding its power grid.

The Landscape of the Gut

The human gastrointestinal tract is approximately nine meters long when stretched out, running from the mouth to the anus and encompassing the esophagus, stomach, small intestine, and large intestine. The small intestine, at roughly six meters in length, is where most nutrient absorption occurs. Its inner surface is carpeted in millions of tiny finger-like projections called villi, and each villus is itself covered in even tinier projections called microvilli, creating a surface area so vast that if the inner lining of the small intestine were unfolded and laid flat, it would cover roughly 30 square meters, the size of a small apartment.

The large intestine, or colon, is shorter but wider, and it is here that the microbiome truly comes into its own. The colon is the most densely microbially inhabited region of the entire human body. As many as 100 billion bacteria can be found in a single milliliter of colonic contents. The colon is their home, their fortress, their ecosystem. Here, in the warm, anaerobic depths of your large intestine, an ecological drama of extraordinary complexity unfolds every moment of every day, as hundreds of bacterial species compete for resources, form alliances, produce metabolites, and collectively perform functions that are essential to your survival.

The Work of Gut Bacteria

The most fundamental job of gut bacteria is fermentation. When food reaches the colon, it has already been partially processed by the stomach's acid and the small intestine's enzymes. What remains is largely the dietary fiber, the complex carbohydrates from plant foods that human digestive enzymes cannot break down. These fibers are the primary food source for colonic bacteria, and the bacteria's fermentation of these fibers produces a range of metabolites that profoundly influence human health. The most important of these metabolites are the short-chain fatty acids: butyrate, acetate, and propionate.

Butyrate deserves special attention. It is the primary energy source for the colonocytes, the cells that line the colon, and it is essential for maintaining their health and integrity. Colonocytes that are deprived of butyrate begin to die, and when they die in sufficient numbers, the gut barrier begins to break down. Butyrate also has powerful anti-inflammatory effects, suppressing the production of inflammatory cytokines and activating regulatory T cells that prevent excessive immune responses. It has been shown to inhibit the growth of colon cancer cells in laboratory studies. It influences the expression of hundreds of genes throughout the body. And it is produced entirely by gut bacteria: the human body cannot make it on its own.

Acetate and propionate are absorbed into the bloodstream and travel to the liver, where they influence metabolic processes including glucose production and fatty acid synthesis. Propionate has been shown to reduce appetite by stimulating the production of satiety hormones. Acetate plays roles in cholesterol metabolism and energy homeostasis. Together, the short-chain fatty acids represent one of the most powerful metabolic connections between the gut microbiome and the rest of the body, a biochemical conversation that influences body weight, metabolic health, cardiovascular risk, and inflammation throughout the entire system.

The Gut as Immune Headquarters

It is a startling fact that the gut houses approximately 70 to 80 percent of the body's entire immune system. The gut-associated lymphoid tissue, or GALT, is the largest immune organ in the human body, containing more immune cells than all the other immune organs combined. This makes ecological sense. The gut is the body's primary interface with the outside world: every day, enormous quantities of food, water, and their associated microbial passengers pass through the digestive system. The immune system needs to be there, ready to respond to genuine threats while simultaneously tolerating the vast quantities of harmless or beneficial microbes and food proteins that pass through.

The gut microbiome is in constant, intimate dialogue with this immune system. Beneficial bacteria produce compounds that signal to immune cells to remain calm, to tolerate dietary antigens and commensal bacteria while remaining alert to genuine pathogens. They train regulatory T cells to suppress inappropriate immune responses. They stimulate the production of IgA antibodies, which coat the gut lining and provide a first line of defense against pathogens. They influence the differentiation of macrophages and dendritic cells, the sentinel cells of the immune system, in ways that tune the overall immune response toward tolerance or defense depending on the circumstances.

When this microbial education is disrupted, when the gut microbiome is depleted or imbalanced, the immune system can go wrong in dramatic ways. Studies in germ-free mice, animals raised from birth in completely sterile conditions with no microbiome at all, have provided some of the most striking evidence for the immune system's dependence on microbial signals. Germ-free mice have profoundly underdeveloped immune systems. Their lymphoid organs are small and poorly organized. Their IgA production is negligible. Their regulatory

T cell populations are minimal. When exposed to normal microbial communities, these deficiencies can be corrected, but the window of opportunity for this correction is limited. Early-life microbial exposure, it appears, is particularly critical for establishing a properly calibrated immune system.

Leaky Gut: When the Border Breaks

The gut lining serves as one of the most important barriers in the human body, a selective gate that allows nutrients to pass into the bloodstream while keeping bacteria, bacterial products, and other potentially harmful substances out. This barrier is formed by a single layer of epithelial cells, connected to each other by tight junction proteins that seal the gaps between them. When this barrier is healthy and intact, it functions remarkably well. When it is compromised, the consequences can be severe and systemic.

Intestinal permeability, commonly known as leaky gut, occurs when the tight junctions between gut epithelial cells become loosened or disrupted, allowing bacterial products, particularly a compound called lipopolysaccharide found on the surface of gram-negative bacteria, to seep through the gut wall and into the bloodstream. Once in the circulation, these bacterial products trigger a low-grade but persistent inflammatory response that has been linked to an astonishing range of chronic diseases including obesity, type 2 diabetes, cardiovascular disease, non-alcoholic fatty liver disease, depression, anxiety, and multiple autoimmune conditions.

The gut microbiome plays a central role in maintaining the integrity of the gut barrier. Certain bacterial species, particularly those that produce butyrate, actively promote the production and maintenance of tight junction proteins. Other bacterial species, and particularly the imbalanced bacterial communities that result from poor diet, antibiotic use, or chronic stress, can damage the gut barrier, contributing to the vicious cycle in which leaky gut leads to inflammation, which further disrupts the microbiome, which further damages the gut barrier.

Germ-Free Mice and the Lessons They Teach

Some of the most revealing experiments in microbiome science involve animals that have none. Germ-free mice are raised from birth in sterile bubbles, eating sterilized food and breathing filtered air, with no microbial exposure whatsoever. These animals provide scientists with a kind of biological blank slate, a mammal without a microbiome, which can then be colonized with specific microbial communities and studied for the effects.

What researchers have found in germ-free mice is extraordinary. These animals have an enormous appetite and yet remain lean; without gut bacteria to ferment dietary fiber and harvest calories from food, they extract far less energy from their meals than normal mice do. Their immune systems are profoundly deficient, as described above. Their intestines are anatomically abnormal, with thin walls, reduced blood supply, and impaired motility. Their stress response is hyperactivated; they show elevated levels of cortisol in response to even mild stressors. And their behavior is abnormal in ways that foreshadow the discoveries we will explore in the chapters on the gut-brain connection.

When germ-free mice are colonized with microbes from healthy normal mice, most of these abnormalities can be corrected. When they are colonized with microbes from diseased or obese donors, they develop the corresponding diseases. These experiments have transformed our understanding of the microbiome from a correlational curiosity into a mechanistic reality. The microbiome does not merely accompany health and disease. In many cases, it causes them.

CHAPTER FIVE

The Immune System and the Microbiome

Your Microbial Army

To understand the relationship between the immune system and the microbiome, you need to let go of a metaphor that has been so deeply embedded in our cultural understanding of medicine that it feels almost like common sense. The metaphor is war. We have been telling ourselves for over a century that the immune system is a military force, an army of specialized cells whose job is to identify and destroy enemy organisms, the bacteria and viruses and fungi that seek to invade and harm us. This metaphor is not entirely wrong. Parts of the immune system really do function like a military force, identifying pathogens, mounting targeted attacks, and producing immunological memory that makes future responses faster and stronger.

But the military metaphor is catastrophically incomplete. It leaves out the most fundamental and, in many ways, most important aspect of the immune system's function: the management of peaceful relations with the trillions of beneficial microbes that the body depends on for its health and survival. The immune system is not simply a war machine. It is a diplomat, a negotiator, a sophisticated manager of biological relationships, calibrating its responses with extraordinary precision to identify which organisms to attack, which to tolerate, and which to actively cultivate. And the microbiome, as we shall see, is not merely a passive recipient of these diplomatic decisions. It is an active participant in shaping them.

The Education of the Immune System

The immune system does not arrive fully formed and perfectly calibrated at birth. It is educated. And the primary educators are the microbes that colonize the gut, the skin, and the other body surfaces during the first months and years of life. This microbial education is not metaphorical; it is a precise, molecular process in which bacterial compounds interact with immune cells in ways that shape their development, their activation thresholds, their tolerance limits, and their long-term functional capacities.

One of the best-understood examples of this microbial education involves a bacterial species called Bacteroides fragilis. This organism, one of the most common inhabitants of the human gut, produces a compound called polysaccharide A on its surface. Research by Caltech immunologist Sarkis Mazmanian has shown that polysaccharide A acts as a direct signal to the developing immune system, stimulating the production and differentiation of regulatory T cells, a class of immune cells that suppress excessive inflammatory responses and maintain immunological tolerance. When germ-free mice are colonized with B. fragilis, their regulatory T cell populations expand and their susceptibility to inflammatory bowel disease decreases dramatically. When they are colonized with a mutant strain of B. fragilis that cannot produce polysaccharide A, the effect disappears. The molecule is the message: the bacterium is speaking directly to the immune system in chemical language.

Similar stories are emerging for dozens of other bacterial species and the immune cell populations they influence. Clostridia bacteria, present in healthy human guts, stimulate the production of anti-inflammatory cytokines that dampen excessive immune reactions. Certain Lactobacillus species tune the responses of dendritic cells, the master coordinators of immune response, toward tolerance rather than attack. The overall picture that emerges from this research is of an immune system that is, in a very real sense, partially constructed from the outside, assembled and calibrated by the microbial community that surrounds it during development.

The Hygiene Hypothesis and Its Evolution

The observation that children raised in clean, sterile environments are more susceptible to allergies and asthma than children raised in environments with more microbial exposure has a long history in medicine. In 1989, British epidemiologist David Strachan proposed what became known as the hygiene hypothesis: the idea that declining rates of childhood infection in industrialized countries, resulting from improved sanitation, antibiotics, and vaccination, were causing the immune system to misdirect its energies toward harmless targets like pollen and pet dander, producing the epidemic of allergic and autoimmune disease that has characterized the late 20th and early 21st centuries.

Strachan's original hypothesis was insightful but imprecise. The problem was not simply that modern children were getting fewer infections. Studies comparing infection rates between industrialized and developing countries did not always support a straightforward relationship between infectious disease burden and allergy rates. The hypothesis needed refinement.

The refinement came in the form of the old friends hypothesis, proposed by immunologist Graham Rook. Rather than infections per se, Rook argued, what the immune system requires for proper development is exposure to the organisms it has co-evolved with over millions of years. These include not just the pathogens that occasionally make us sick but the whole community of environmental bacteria, helminths, and other microorganisms that were a constant presence in the ancestral human environment. These are the immune system's old friends, organisms it has co-existed with for so long that it has learned to recognize them as non-threatening and to respond to their presence with regulatory rather than inflammatory signals. Remove these old friends from the environment, as modern sanitation and hygiene have done, and the immune system, calibrated for a world full of microbial encounters, becomes poorly regulated, prone to false alarms and excessive responses.

Autoimmune Disease: When the System Turns on Itself

The consequences of a poorly calibrated immune system are not limited to hay fever and nut allergies. Autoimmune diseases, conditions in which the immune system mistakenly attacks the body's own tissues, have been increasing in prevalence across all industrialized nations for decades. Rheumatoid arthritis, multiple sclerosis, type 1 diabetes, lupus, inflammatory bowel disease: these conditions affect hundreds of millions of people worldwide and cause enormous suffering. And the evidence linking them to disruptions of the microbiome is growing stronger with every passing year of research.

Multiple studies have found that people with autoimmune diseases consistently show patterns of gut dysbiosis, alterations in the composition and diversity of their gut microbiome, that distinguish them from healthy individuals. In some cases, these microbial differences precede the onset of disease, suggesting that the dysbiosis may be contributing to the development of autoimmunity rather than simply resulting from it. In animal models, the evidence is more direct: germ-free mice do not develop many of the autoimmune conditions that affect normal mice. When they are colonized with specific bacterial communities, some of those conditions can be induced or prevented, depending on which microbes are introduced.

Particularly striking is the case of multiple sclerosis, a devastating autoimmune disease in which the immune system attacks the myelin sheath that protects nerve fibers in the brain and spinal cord. Researchers have found consistent differences in the gut microbiomes of MS patients compared to healthy controls, with MS patients showing reduced abundance of certain anti-inflammatory bacterial species. When gut bacteria from MS patients are transferred to germ-free mice, those mice develop more severe inflammatory responses than mice that receive bacteria from healthy donors. The gut microbiome, residing far from the brain, is influencing a disease that destroys brain tissue. The connection is mediated by the immune system, which is shaped by the microbiome and then shapes the neurological environment. The distance between the gut and the brain, it turns out, is much shorter than anatomy suggests.

The Antibiotic Paradox

Antibiotics are among the most consequential medical inventions in human history. Before their discovery, bacterial infections that we now treat as minor inconveniences, strep throat, urinary tract infections, pneumonia, were routinely fatal. The introduction of penicillin in the 1940s, followed by decades of antibiotic development, transformed medicine and saved hundreds of millions of lives. No serious discussion of antibiotics can ignore this enormous benefit.

But antibiotics do not discriminate. They kill bacteria, and they kill not just the pathogenic bacteria that cause the infection being treated but also vast numbers of the beneficial bacteria that constitute the gut microbiome. A single course of a broad-spectrum antibiotic can eliminate hundreds of bacterial species from the gut, dramatically reducing its diversity and disrupting the ecological balance that supports normal immune function. In most adults with otherwise healthy microbiomes, the community recovers over a period of weeks to months, though some studies suggest that certain species may never fully return after antibiotic exposure.

The consequences of early childhood antibiotic exposure are particularly concerning. Studies have consistently found associations between antibiotic use in the first years of life and increased risk of asthma, allergies, inflammatory bowel disease, and obesity in later childhood. These associations do not prove causation, but they are consistent with a mechanistic hypothesis: that early antibiotic treatment, occurring during the critical window of immune system development and microbiome establishment, disrupts the microbial education of the immune system in ways that increase vulnerability to these conditions. The antibiotic paradox is that the very medicines that protect against infection may, when overused, compromise the deeper ecological foundations of immune health.

CHAPTER SIX

The Microbiome, Metabolism, and the Obesity Epidemic

The Hidden Driver of Weight

In 2006, a team of scientists at Washington University in St. Louis conducted an experiment that should have fundamentally changed the way the entire world thinks about weight. The scientists, led by microbiologist Jeffrey Gordon, took two groups of germ-free mice and colonized them with different microbial communities. One group received gut bacteria taken from obese mice. The other received gut bacteria from lean mice. Both groups ate the same diet, in the same amounts, under identical conditions. Within two weeks, the mice that had received the obese microbiome had gained significantly more body fat than those that had received the lean microbiome, despite consuming the same number of calories. The conclusion was unavoidable and revolutionary: the gut microbiome influences body fat accumulation independently of diet.

This experiment, and the flood of research that followed it, has forced a fundamental reassessment of obesity, one of the most pressing public health crises of our time. For decades, the dominant framework for understanding obesity has been the energy balance model: weight gain occurs when calorie intake exceeds calorie expenditure, full stop. Within this framework, obesity is understood as primarily a failure of willpower and discipline, a personal moral failing dressed up in the language of metabolism. This framework is not entirely wrong. Calorie balance does matter. But it is spectacularly incomplete, and nowhere is its incompleteness more apparent than in the science of the microbiome.

How Microbes Harvest Energy

The gut microbiome's ability to influence body weight operates through several distinct mechanisms, each of which represents a point of departure from the simple calories-in, calories-out model. The first and perhaps most fundamental of these mechanisms is caloric extraction: the ability of gut bacteria to harvest additional energy from food that the human digestive system would otherwise excrete.

When dietary fiber reaches the colon, human digestive enzymes are unable to break it down. Left to its own devices, the human body would extract very little energy from fiber and excrete most of it in the stool. But gut bacteria can ferment fiber, breaking it down into short-chain fatty acids that are absorbed through the colon wall and used as fuel by the body's cells. This bacterial fermentation process effectively extracts additional calories from food that would otherwise be wasted. The amount of additional energy extracted through this process varies significantly depending on the composition of the individual's gut microbiome. Some microbial communities are much more efficient energy harvesters than others, meaning that for the same food intake, a person with a highly efficient gut microbiome may absorb significantly more calories than someone with a less efficient one.

This finding has profound implications for our understanding of obesity. It suggests that two people can eat exactly the same meal and, depending on the composition of their gut microbiomes, absorb meaningfully different amounts of energy from it. The person who absorbs more will, over time, be at greater risk of weight gain, not because they are eating more or moving less, but because their gut bacteria are more efficient at extracting energy from their food.

Hormones, Hunger, and the Microbial Conductor

Beyond caloric extraction, the gut microbiome exerts profound control over the hormonal signals that regulate appetite, satiety, and metabolism. The gut is the largest endocrine organ in the human body, producing more than 30 different hormones that influence everything from digestion to blood sugar regulation to hunger. And the production of many of these hormones is directly influenced by the gut microbiome.

Ghrelin, often called the hunger hormone, is produced in the stomach and rises sharply before meals, triggering appetite. Leptin, produced by fat cells, signals to the brain that the body has adequate energy stores and suppresses appetite. GLP-1 and PYY, produced in the gut in response to eating, signal satiety and slow the rate of gastric emptying. Research has consistently shown that the gut microbiome influences the production and regulation of all of these hormones. Specific bacterial metabolites, including short-chain fatty acids, stimulate the cells of the gut lining to produce GLP-1 and PYY, promoting satiety. Other microbial products influence insulin sensitivity, the body's ability to respond to the blood sugar-regulating hormone insulin, in ways that can either protect against or promote the development of type 2 diabetes.

The Firmicutes-Bacteroidetes Ratio

One of the most widely discussed findings in the field of microbiome and obesity research is the observation that obese individuals tend to have a different ratio of the two dominant bacterial phyla in the gut, the Firmicutes and the Bacteroidetes, compared to lean individuals. Specifically, obese individuals tend to have a higher proportion of Firmicutes and a lower proportion of Bacteroidetes. When obese individuals lose weight through dietary intervention, their microbiome shifts back toward the lean pattern, with Bacteroidetes increasing relative to Firmicutes.

This finding has been enormously influential and has sometimes been overstated in popular discussions of the microbiome. It is important to understand what it does and does not mean. It does not mean that simply increasing Bacteroidetes will cause weight loss, or that having more Firmicutes inevitably causes obesity. The relationship is more complex and more bidirectional than that. Many Firmicutes species are entirely beneficial, and some Bacteroidetes species can be harmful. The ratio is a correlational marker, not a causal mechanism in itself. What the research reveals is not a simple good bacterium versus bad bacteria story, but rather a complex ecological shift that accompanies metabolic dysfunction and that may both reflect and contribute to that dysfunction.

Artificial Sweeteners: The Paradox of the Zero-Calorie Option

One of the most counterintuitive findings to emerge from microbiome research concerns artificial sweeteners, those supposedly safe sugar substitutes that have been promoted for decades as a tool for weight management. The logic seemed impeccable: artificial sweeteners provide sweetness without calories, so they should help people consume fewer calories and lose weight. For years, this logic was accepted largely without question. The microbiome research has complicated it significantly.

A landmark study published in the journal Nature in 2014 by Eran Segal, Eran Elinav, and colleagues at the Weizmann Institute of Science found that commonly used artificial sweeteners, including saccharin, sucralose, and aspartame, caused significant changes in the gut microbiomes of mice, and that these changes were associated with glucose intolerance, the very metabolic disruption that artificial sweeteners were supposed to help prevent. When the researchers transferred the gut bacteria from sweetener-treated mice to germ-free mice, those recipient mice also developed glucose intolerance, confirming that the microbial changes, not some direct metabolic effect of the sweeteners themselves, were responsible for the effect. A follow-up study in human volunteers confirmed that about half of them developed significant glucose intolerance after consuming artificial sweeteners for just one week, with the effect mediated by changes in their gut microbiomes.

The Future of Obesity Medicine

The implications of all this research for the treatment of obesity are profound and largely have not yet been incorporated into mainstream clinical practice. If the gut microbiome is a major determinant of metabolic efficiency, appetite regulation, and fat storage, then treating obesity without addressing the microbiome is like trying to fix a malfunctioning car by only adjusting the speedometer. You might change the reading, but you are not addressing the underlying problem.

The next generation of obesity medicine will need to be microbiome-aware. This does not mean that diet and physical activity are irrelevant; they remain crucial, and they are also, as we have seen, powerful tools for reshaping the microbiome. But it does mean that personalized approaches, tailored to the specific microbial landscape of each individual, are likely to be far more effective than the one-size-fits-all dietary prescriptions that have dominated weight management advice for decades. It means that the development of targeted microbial therapies, specific probiotic formulations or dietary interventions designed to shift the microbiome toward a lean-associated profile, represents one of the most promising frontiers in metabolic medicine. And it means that we need to stop blaming people for their weight and start understanding the invisible biological forces that are shaping it.

PART THREE

THE GUT-BRAIN AXIS

The Most Astonishing Connection in Human Biology

CHAPTER SEVEN

The Second Brain

Your Gut Has a Mind of Its Own

There is a conversation happening inside your body right now. It has been happening every moment of your life, from before you were born until this very second. It is a conversation conducted not in words but in the language of electricity and chemistry, in the rapid-fire signals of neurons and the slower, more deliberate vocabulary of hormones and neurotransmitters. What is extraordinary is where this conversation is happening. Not in the brain inside your skull. In the brain inside your belly.

The enteric nervous system, the collection of neurons that lines your gastrointestinal tract from the esophagus to the rectum, contains between 200 and 600 million nerve cells. To put that number in context: that is more neurons than are found in the spinal cord. More than in many regions of the brain itself. This enteric nervous system is so complex, so independent, and so functionally autonomous that neuroscientists have given it its own name: the second brain. And like the brain in your head, it can sense, process, and respond to information entirely on its own, without instructions from above.

A Nervous System That Evolved First

From an evolutionary perspective, the enteric nervous system may actually be older than the central nervous system. In early multicellular animals, the gut was the primary interface with the environment, the organ through which nutrients were acquired and pathogens were resisted. A nervous system to manage this critical organ was, from an evolutionary standpoint, a more immediate priority than a centralized brain for higher cognition. The enteric nervous system shows deep evolutionary conservation: the basic architecture of the gut nervous system in nematode worms is strikingly similar to that found in humans, separated by hundreds of millions of years of evolution. This ancient lineage is part of what makes it so sophisticated.

The enteric nervous system contains not only neurons but also glial cells, the supporting cells of neural tissue, in proportions similar to those found in the brain. It has its own reflexes, its own neurotransmitters, its own sensory, motor, and interneuronal circuitry. It can coordinate the complex muscular contractions of peristalsis, the wave-like movements that propel food through the gut, entirely independently of the brain. People who have suffered spinal cord injuries that sever all communication between the brain and the gut still have functioning digestive systems, because the enteric nervous system handles the job on its own.

The Serotonin Surprise

If you asked most people where serotonin is produced in the body, the overwhelming majority would say the brain. Serotonin is, after all, the neurotransmitter most associated with mood, happiness, and emotional well-being. Antidepressant drugs that boost serotonin levels in the brain are among the most widely prescribed medications in the world. Surely serotonin is a brain chemical, produced in the brain and acting on the brain.

The reality is one of the most surprising facts in all of neuroscience. Approximately 90 to 95 percent of all the serotonin in the human body is produced not in the brain but in the gut, specifically in the enterochromaffin cells lining the intestinal wall. The gut, not the brain, is the body's primary serotonin factory. Gut bacteria play a crucial role in this production: research has shown that germ-free mice have dramatically lower levels of gut serotonin than normal mice, and that colonization with specific bacterial species can restore serotonin production. In other words, without the right gut bacteria, you cannot produce normal amounts of serotonin. The implications for mental health are enormous and still being worked out.

Gut serotonin does not function exactly like brain serotonin. It is largely confined to the gut environment, where it regulates gastrointestinal motility, secretion, and sensitivity to pain. But it also influences the vagus nerve, the major communication pathway between the gut and the brain, and this is where the story becomes deeply relevant to understanding mood, anxiety, and mental health.

The Vagus Nerve: The Superhighway Between Worlds

The vagus nerve is the longest cranial nerve in the human body. It originates in the brainstem, travels down through the neck and chest, and branches into the heart, lungs, and digestive organs before terminating in the colon. Its name comes from the Latin word for wandering, an apt description of its circuitous journey through the body. The vagus nerve is the primary physical connection between the gut and the brain, and it is the most important highway of the gut-brain axis.

Here is the fact that surprises most people when they first encounter it: approximately 80 percent of the signals carried by the vagus nerve travel from the gut to the brain, not the other way around. The gut, it turns out, talks to the brain far more than the brain talks to the gut. The gut is sending a continuous stream of sensory information upward: information about the chemical composition of its contents, the state of its microbial community, the degree of mechanical stretch in its walls, the presence of pathogens or toxins. The brain uses this information to regulate appetite, mood, stress responses, and many other functions. But it is fundamentally reactive to information coming from below, not simply commanding from above.

The role of gut bacteria in this communication is central. Bacteria in the gut produce a range of compounds that stimulate the sensory nerve endings of the vagus nerve, sending signals to the brain. Short-chain fatty acids produced by bacterial fermentation stimulate specialized cells in the gut lining called enterochromaffin cells, which in turn release serotonin and other signaling molecules that activate vagal afferent neurons. GABA, another crucial neurotransmitter, is produced directly by certain gut bacteria and influences vagal signaling. The gut microbiome is, in this sense, not merely a passive resident of the digestive system but an active participant in the neurological conversation between gut and brain.

Gut Feelings Are Real

The phrase gut feeling is not merely a figure of speech. It reflects a genuine biological reality: the gut is a sensory organ, capable of detecting and processing information about the internal environment and sending that information to the brain in ways that influence conscious experience. When you feel a knot of anxiety in your stomach before an important presentation, when you feel the warm glow of satisfaction after a good meal, when you feel nausea in response to something morally disgusting, these are not just metaphors translated into physical sensation. They are real neurobiological events, arising from the enteric nervous system and communicated to the brain via the vagus nerve.

Research by neuroscientist A.D. Craig and others has revealed that the brain's interoceptive processing, its reading of internal body states, is deeply integrated with emotional processing. The insula, a brain region central to the experience of emotion, receives and processes signals from the body's internal organs, including the gut. Gut signals contribute to the brain's moment-to-moment assessment of the body's state, and this assessment is a fundamental component of what we experience as emotion and intuition. When we talk about following our gut, we are describing, in colloquial language, a process that is genuinely neurobiological: the brain integrating signals from the enteric nervous system into its overall assessment of a situation.

CHAPTER EIGHT

Microbes and the Mind

How Gut Bacteria Shape Your Mood, Anxiety, and Mental Health

"If we accept that the gut microbiome can influence the brain, then we must also accept that our sense of self, our moods, our anxieties, our very character, are partly the product of the organisms we carry within us."

— Prof. John Cryan, Neuropharmacologist

The first time Ted Dinan and John Cryan proposed, in a serious scientific forum, that it might be possible to treat depression and anxiety by targeting the gut microbiome, the response from the psychiatric community was somewhere between polite skepticism and outright disbelief. Depression is a brain disease, the conventional wisdom held. Its causes are neurological. Its treatments are pharmacological or psychological. The idea that the solution might lie in the gut, in the bacteria fermenting your dietary fiber, was the kind of notion that could end careers if stated too loudly in the wrong rooms.

That was less than twenty years ago. Today, the field of psychobiotics, the study of how gut microorganisms influence brain function and mental health, is one of the fastest growing areas in all of neuroscience. The evidence that has accumulated in that time is not merely suggestive. In many domains, it is striking. And it is forcing a fundamental reconsideration of what mental illness is, where it comes from, and how it might best be treated.

The Psychobiome: Your Mental Microbiota

Scientists have coined the term psychobiome to refer to the subset of the gut microbiome with particular relevance to brain function and mental health. The psychobiome operates through multiple overlapping channels. Gut bacteria produce or stimulate the production of neurotransmitters and their precursors, including serotonin, dopamine, GABA, and norepinephrine. They produce short-chain fatty acids that cross the blood-brain barrier and influence neuronal function. They modulate the stress response through their effects on the hypothalamic-pituitary-adrenal axis, the hormonal system that controls the production of cortisol. They influence neuroinflammation, the inflammatory processes occurring within the brain and central nervous system, which are increasingly recognized as playing a major role in depression, anxiety, and other psychiatric conditions.

GABA, gamma-aminobutyric acid, is the brain's primary inhibitory neurotransmitter. It calms neural activity, reduces anxiety, and promotes relaxation and sleep. Benzodiazepine drugs like Valium work by enhancing the effects of GABA in the brain, which is why they reduce anxiety and promote sleep. In 2011, John Cryan's team at University College Cork published a landmark study showing that Lactobacillus rhamnosus, a bacterial species common in the gut, produces GABA directly and that feeding mice this bacterium significantly reduced anxiety-related behavior and altered the expression of GABA receptor genes in the brain. Crucially, when the vagus nerve was severed in these animals, the effect on brain chemistry and behavior disappeared completely, confirming that the gut bacteria were communicating with the brain via the vagal pathway.

The Microbiome of Depression

Major depressive disorder affects more than 300 million people worldwide. It is the leading cause of disability globally and is associated with enormous personal suffering and economic cost. Antidepressant medications help many patients, but at least one-third of depressed patients do not respond adequately to currently available treatments. The search for better therapeutic targets has driven researchers increasingly toward the microbiome.

Multiple studies across different countries and populations have found that people with major depressive disorder have consistent differences in their gut microbiome composition compared to healthy controls. Depressed individuals typically show reduced abundance of several genera associated with health and anti-inflammatory function, including Lactobacillus, Bifidobacterium, and Faecalibacterium prausnitzii, which is one of the most abundant and important anti-inflammatory bacteria in the healthy gut. They show increased abundance of certain pro-inflammatory species. These differences are not small or inconsistent. They are robust enough that machine learning algorithms trained on microbiome data can classify patients as depressed or not depressed with meaningful accuracy.

But is the altered microbiome a cause of depression, or merely a consequence of the lifestyle changes, dietary shifts, and physiological stress that accompany it? This is the crucial question, and the most striking evidence bearing on it comes from transplant experiments. In 2019, researchers published a study in which they transferred fecal samples from people with major depressive disorder into germ-free rats. These rats, who had never been exposed to any microbes and therefore had no preexisting behavioral phenotype, subsequently developed behaviors characteristic of depression in rodents: they showed reduced motivation, altered stress responses, and changes in tryptophan metabolism, a key biochemical pathway in depression. Rats that received fecal transplants from healthy human donors did not develop these behaviors. The depressed microbiome, transferred from humans to rats, caused depressive-like behavior in animals that had never had a microbiome before. This is powerful causal evidence.

Anxiety, the Microbiome, and the Stress Response

Anxiety disorders are even more prevalent than depression, affecting an estimated 284 million people globally. The relationship between the gut microbiome and anxiety is one of the best characterized in the field of psychobiotics, partly because anxiety can be studied relatively easily in animal models where the microbiome can be experimentally manipulated.

Germ-free mice, as described in earlier chapters, exhibit dramatically elevated anxiety compared to normal mice. They show increased activity in the hypothalamic-pituitary-adrenal axis, producing more corticosterone (the mouse equivalent of cortisol) in response to stressors. They show exaggerated anxiety-like behavior in standard behavioral tests. And remarkably, when germ-free mice are colonized with bacteria from calm, low-anxiety mouse strains, their anxiety decreases. When they are colonized with bacteria from high-anxiety mouse strains, it increases. The microbiome, transplanted from one animal to another, transfers the anxiety phenotype.

In human studies, probiotic interventions targeting anxiety have shown promising results. A systematic review of randomized controlled trials published in 2019 found that probiotic supplementation produced statistically significant reductions in anxiety scores across multiple studies, with Lactobacillus and Bifidobacterium species showing the most consistent effects. These effects were modest compared to pharmacological treatments, but they were real, and they were achieved with interventions that have essentially no side effects, a significant advantage over benzodiazepines, which carry substantial risks of dependence and cognitive impairment.

The Autism Connection

Few areas of psychobiome research are more sensitive or more controversial than the connection between the gut microbiome and autism spectrum disorder. Autism is a complex, heterogeneous condition that affects approximately 1 in 44 children in the United States and is characterized by differences in social communication, restricted and repetitive behaviors, and sensory processing. Its causes are not fully understood, but the evidence for significant genetic contributions is strong.

The gut connection to autism is, however, impossible to ignore. Gastrointestinal symptoms are among the most common co-occurring conditions in autism, affecting between 46 and 84 percent of autistic individuals in various studies. Children with autism consistently show different gut microbiome compositions than neurotypical children. Specific bacterial species are altered: Prevotella, a genus associated with a healthy, plant-rich diet, is consistently found at lower abundance. Certain Clostridium species, known to produce toxic metabolites, are elevated. Bifidobacterium, crucial in early life for gut barrier integrity and immune development, is often reduced.

Whether these microbiome differences contribute to autistic traits or are merely a consequence of the dietary selectivity common in autism is a subject of active debate. Animal research suggests a causative role: studies in mice have shown that certain microbial alterations can produce social behavior deficits, and that restoration of the microbiome can partially reverse these deficits. Probiotic interventions in small human trials have produced improvements in gastrointestinal symptoms and, in some cases, behavioral measures. The research is preliminary, and the field is rightly cautious about overstating the evidence. But the gut-brain connection in autism is real, biologically plausible, and worthy of serious scientific attention.

CHAPTER NINE

Stress, Sleep, and the Microbiome

The Vicious Cycle

Consider what happens to a forest during a drought. The trees do not disappear overnight. But under sustained stress, the delicate ecological balance of the forest ecosystem begins to shift. Drought-tolerant species gain ground. Species that depend on moisture retreat. Fungi that were once partners of the trees become competitive or parasitic. Insects that were kept in check by healthy predator-prey dynamics begin to proliferate unchecked. The forest looks similar to before, still green, still standing, but its inner ecology has shifted in ways that increase its vulnerability to fire, to disease, to collapse. The shift may be invisible to a casual observer. Its consequences are not.

This is a useful metaphor for what happens to the human gut microbiome under chronic stress. The community does not vanish. But its composition changes, its diversity decreases, its functional balance tips, and its resilience to further disturbance diminishes. And unlike a forest, which at least experiences the drought from the outside, the microbiome is being stressed from within, by the very host it depends on for its survival. The relationship between stress, sleep, and the microbiome is not a simple cause-and-effect story. It is a loop, a feedback cycle in which stress damages the microbiome, the damaged microbiome amplifies the stress response, the amplified stress disrupts sleep, the disrupted sleep further damages the microbiome, and so on, round and round, potentially for years.

Cortisol and the Microbiome: A Toxic Relationship

When the body perceives a threat, real or imagined, the hypothalamic-pituitary-adrenal axis activates the stress response. The hypothalamus signals the pituitary gland, which signals the adrenal glands to release cortisol, the body's primary stress hormone. Cortisol mobilizes energy, sharpens attention, suppresses non-essential functions, and prepares the body for either fight or flight. In acute, time-limited situations, this response is adaptive and life-saving. The problem arises when stress is chronic and the cortisol response never fully turns off.

Chronic elevation of cortisol has direct and measurable effects on the gut microbiome. Cortisol alters the motility of the gut, changing how quickly contents move through the digestive system. It affects the secretion of mucus that lines the gut wall, a protective layer that also serves as habitat for many beneficial bacterial species. It alters the permeability of the gut lining, contributing to the leaky gut phenomenon described in earlier chapters. It directly changes the composition of the gut microbial community, reducing the abundance of beneficial species and creating conditions more favorable to opportunistic pathogens.

Research in both animals and humans has confirmed these effects. Animal studies using various stress models, including social defeat stress, early life separation stress, and restraint stress, consistently find significant alterations in gut microbiome composition following stress exposure. Human studies in medical students during examination periods, in people experiencing work-related burnout, and in populations affected by natural disasters have found corresponding microbial changes. The stressed gut is a less diverse, less resilient, less functionally competent gut, and the downstream consequences for immune function, metabolism, and brain chemistry are significant.

Early Life Stress: Wounds That Persist

Some of the most profound effects of stress on the microbiome occur during early childhood, when the microbial community is still being established and when the developing organism is particularly sensitive to environmental perturbations. Research on early life adversity, including childhood abuse, neglect, poverty, and household dysfunction, has found that these experiences leave lasting marks on the gut microbiome that can persist for decades.

Studies of children who experienced early institutional care, such as those raised in understaffed orphanages before being adopted into family environments, show significantly reduced microbiome diversity compared to children raised in family settings from birth. The reduced diversity associated with early adversity is correlated with higher rates of anxiety, depression, and immune dysregulation in adulthood. Research in animal models has gone further, showing that early maternal separation, a well-established animal model of early life stress, produces gut microbiome changes that alter brain development, increase anxiety and depressive behavior, and raise the baseline stress response in ways that persist throughout the animal's lifetime.

These findings are deeply relevant to public health. The understanding that early childhood adversity leaves biological marks not just in the brain but in the gut, microbiome adds a new dimension to the science of adverse childhood experiences. It suggests that interventions targeting the microbiome might play a role in mitigating the long-term health consequences of early trauma, a possibility that is being actively explored in clinical research.

The Microbiome and Your Sleep

Sleep is one of the most fundamental biological requirements of the human organism. In the hierarchy of physiological needs, it rivals food and water in its essentiality. We know this not just abstractly but from painful personal experience: the fog of cognitive impairment after a sleepless night, the emotional volatility that follows days of inadequate rest, the physical sense of systemic malfunction that comes with chronic sleep deprivation. What is less widely known is that the gut microbiome and sleep are intimately connected, each influencing the other through mechanisms that are only beginning to be understood.

Gut bacteria follow their own circadian rhythms, synchronized with the host's sleep-wake cycle. The composition and abundance of different bacterial species fluctuate over the course of the day in a regular, predictable pattern that mirrors the host's daily biological rhythms. These microbial circadian rhythms are coordinated by the host's circadian clock, the molecular mechanism that controls the body's 24-hour biological cycles. When the circadian clock is disrupted, as it is in shift workers, people experiencing jet lag, and individuals with insomnia, the microbiome's rhythms become desynchronized, and the resulting microbial dysfunction contributes to the metabolic and immune problems associated with circadian disruption.

Conversely, the microbiome produces compounds that directly influence sleep. Tryptophan, an amino acid that gut bacteria help regulate, is the precursor to both serotonin and melatonin, the hormone that regulates sleep timing. Gut bacteria influence the metabolism of tryptophan in ways that affect how much reaches the brain and how much is converted to melatonin versus other products. Short-chain fatty acids produced by bacterial fermentation have been shown to influence the expression of circadian clock genes in the liver and potentially in the brain. The microbiome is not just influenced by sleep; it actively participates in regulating it.

Breaking the Cycle

The relationship between stress, sleep, and the microbiome is a loop, but it is not a closed one. It can be entered and disrupted at multiple points. Research has shown that interventions targeting the microbiome can reduce stress reactivity and improve sleep quality. Probiotic supplementation with Lactobacillus and Bifidobacterium strains has been found in several randomized controlled trials to reduce cortisol levels, improve scores on anxiety and depression measures, and improve both subjective sleep quality and objective sleep parameters in adults with stress-related sleep problems.

Dietary interventions have similar effects. A randomized trial published in 2021 by researchers at Stanford found that a diet high in fermented foods significantly increased microbiome diversity and reduced markers of inflammatory cytokines, including those associated with stress and anxiety. The Mediterranean diet, consistently associated with better mental health outcomes in epidemiological studies, is also the diet most consistently associated with a healthy, diverse gut microbiome. The connection is not coincidental. The diet feeds the microbiome. The microbiome regulates the stress response. The regulated stress response improves sleep. The improved sleep supports a healthy microbiome. The virtuous cycle, once established, reinforces itself. Getting there requires breaking the vicious one, and that is the subject of the book's final chapters.

PART FOUR

THE MICROBIOME AND DISEASE

When the Inner Universe Collapses

CHAPTER TEN

The Microbiome and Cancer

An Unexpected Connection

In the oncology clinic, certainties are rare and carefully guarded. Cancer is a field accustomed to humility. Treatments that seemed revolutionary often prove effective only for some patients, in some tumor types, under some circumstances that take years of research to fully delineate. The history of oncology is as much a history of promising leads that turned into dead ends as it is a history of genuine breakthroughs. So when, in 2018, three independent research teams published papers in the prestigious journals Science and Nature showing that the efficacy of the most exciting new class of cancer drugs was dramatically influenced by the patient's gut microbiome, the oncology world paid close attention.

The drugs in question were immune checkpoint inhibitors, a class of immunotherapy that had transformed cancer treatment by unleashing the immune system's natural capacity to recognize and destroy cancer cells. Checkpoint inhibitors work by blocking the proteins that cancer cells use to disguise themselves from immune surveillance. When they work, they can produce dramatic and durable remissions in cancers that had previously been considered untreatable. But they do not work for everyone. In some tumor types, fewer than 20 percent of patients respond. The oncology community had long sought to understand why, and the answer, it turned out, was at least partly to be found in the gut.

The Tumor Microenvironment and Microbial Influence

Cancer does not grow in isolation. Tumors are embedded in a complex ecosystem of immune cells, stromal cells, blood vessels, and, as researchers are now discovering, microorganisms. The tumor microenvironment, the immediate cellular and molecular context in which a tumor grows, is now understood to be a critical determinant of tumor behavior and treatment response. The immune cells present in the tumor microenvironment can either suppress tumor growth, if they are properly activated and appropriately targeting the cancer, or inadvertently support it, if they are suppressed or misdirected.

The gut microbiome influences the tumor microenvironment through multiple mechanisms. It shapes the overall immune landscape of the body, determining the baseline activation state and functional capacity of the immune cells that infiltrate tumors. It produces metabolites that circulate in the bloodstream and influence immune cell differentiation at sites distant from the gut. And, in some cancers, it appears that microorganisms may be present within the tumor itself, directly influencing the local immune environment.

The 2018 studies found that patients whose gut microbiomes contained certain bacterial species, particularly Akkermansia muciniphila, Faecalibacterium prausnitzii, and Bifidobacterium species, responded significantly better to immune checkpoint inhibitors than patients whose microbiomes lacked these organisms. In one particularly striking demonstration, researchers transferred fecal samples from responding patients into germ-free mice with human cancer cell tumors. The mice that received the microbiomes from treatment responders showed improved immune activation against their tumors compared to mice that received microbiomes from non-responders. The gut bacteria were modulating the anti-tumor immune response in a way that determined whether the treatment succeeded or failed.

Fecal Transplants for Cancer

The logical extension of these findings is to ask whether transferring a favorable gut microbiome to a patient who is not responding to cancer immunotherapy might improve their outcome. This is exactly what several clinical research groups have now attempted, with early results that have generated considerable excitement in the oncology community.

In 2021, two papers published in Science reported the results of small clinical trials in which patients with melanoma who had failed to respond to immune checkpoint inhibitors received fecal microbiota transplants from patients who had achieved excellent responses to those drugs. In both trials, a subset of the non-responding patients developed favorable gut microbiome changes following the transplant, and some of these patients went on to experience objective clinical benefit, including tumor shrinkage, when re-challenged with the checkpoint inhibitor that had previously failed them. The results were preliminary and involved small numbers of patients, but they provided proof-of-concept that microbiome modulation might be a new tool in the cancer treatment arsenal.

Bacteria Inside Tumors

One of the most surprising recent discoveries in cancer biology is the finding that tumors themselves harbor microbial communities. This was not anticipated. The interior of a solid tumor, with its disrupted blood supply, hypoxic environment, and chaotic cellular architecture, was not expected to be a hospitable environment for bacteria. Yet advanced sequencing techniques have revealed that many tumor types, including colorectal cancer, pancreatic cancer, breast cancer, and lung cancer, contain detectable bacterial populations.

The relationship between intratumoral bacteria and tumor biology is complex and still being worked out. In colorectal cancer, a particularly well-studied case, the bacterium Fusobacterium nucleatum has been found to be enriched within tumor tissue relative to adjacent normal colon tissue. F. nucleatum appears to promote cancer progression through several mechanisms: it activates cancer cell signaling pathways that promote proliferation, it suppresses anti-tumor immune responses within the tumor microenvironment, and it may contribute to resistance to chemotherapy. Crucially, colorectal cancer patients with high intratumoral F. nucleatum abundance have consistently worse clinical outcomes than those with low abundance, suggesting that this bacterium is not merely an innocent bystander but an active participant in the biology of the tumor.

Microbiome-Based Cancer Prevention

Beyond cancer treatment, the microbiome may also offer opportunities for cancer prevention. The gut microbiome influences carcinogenesis, the process by which normal cells transform into cancer cells, through its effects on inflammation, DNA damage, and epithelial cell proliferation. Chronic intestinal inflammation, often driven by a dysbiotic microbiome, is a well-established risk factor for colorectal cancer. Bacterial metabolites can either promote or protect against DNA damage in gut epithelial cells, depending on the composition of the microbial community. A diverse, health-promoting gut microbiome, characterized by high abundance of short-chain fatty acid producers and low abundance of pro-inflammatory species, may confer protection against colorectal cancer risk.

The evidence on this point is consistent with the known cancer-protective effects of a diet rich in plant fiber and fermented foods, both of which promote a diverse and beneficial gut microbiome. The Mediterranean diet, high in fiber, polyphenols, and fermented dairy and vegetable products, is consistently associated with reduced colorectal cancer risk in epidemiological studies. Part of this protection almost certainly operates through the microbiome, which is shaped by this diet in ways that reduce intestinal inflammation, decrease the production of carcinogenic metabolites, and maintain the integrity of the gut epithelium that serves as the front line against colorectal carcinogenesis.

CHAPTER ELEVEN

Antibiotics

The Double-Edged Sword That Changed the World

September 3, 1928. A Scottish bacteriologist named Alexander Fleming returns to his laboratory at St. Mary's Hospital in London after a two-week holiday. He had left some bacterial cultures growing in Petri dishes on his lab bench. As he examines them, he notices something unusual: one of the dishes, which had been accidentally contaminated with a mold, shows a ring of clearing around the mold colony. The bacteria that had been growing densely across the dish have been killed or inhibited in the region closest to the mold. Fleming studies the mold. It is Penicillium notatum. He deduces that it is producing some substance that kills bacteria. He names this substance penicillin.

Fleming's discovery, and the subsequent work of Howard Florey, Ernst Boris Chain, and their colleagues to isolate and mass-produce penicillin, stands as one of the transformative moments in the history of medicine. Before antibiotics, bacterial infections that we now treat with a brief course of oral medication were routinely fatal. Streptococcal throat infections, pneumonia, bacterial meningitis, wound infections: these were killers. The introduction of penicillin and subsequent antibiotics saved hundreds of millions of lives and fundamentally transformed the relationship between humanity and infectious disease. No serious discussion of antibiotics can begin without this acknowledgment: they are among the most beneficial interventions medicine has ever developed.

How Antibiotics Work

Antibiotics kill or inhibit the growth of bacteria through a variety of mechanisms, each targeting a feature of bacterial biology that differs sufficiently from mammalian biology to allow selective toxicity: the killing of bacteria without killing the patient. Beta-lactam antibiotics like penicillin and its descendants target the enzymes that bacteria use to build their cell walls. Without cell wall synthesis, dividing bacteria cannot maintain structural integrity and lyse, rupturing and dying. Macrolide antibiotics like erythromycin target bacterial ribosomes, the molecular machines that translate genetic code into proteins, interfering with protein synthesis in ways that either kill the bacteria or prevent their growth. Fluoroquinolone antibiotics target bacterial DNA gyrase and topoisomerase, enzymes essential for DNA replication and repair. Without these enzymes, bacteria cannot copy their genomes and die.

All of these mechanisms target bacterial-specific structures, but none of them is perfectly selective. The gut microbiome contains hundreds of species of bacteria, the vast majority of which are beneficial or harmless. When a patient takes a broad-spectrum antibiotic for, say, a urinary tract infection caused by E. coli, the drug travels through the gut on its way to the infection site and encounters billions of innocent bystander bacteria. It kills them too. The collateral damage is massive, swift, and in some cases surprisingly durable.

The Ecological Aftermath

Studies examining the gut microbiome before and after antibiotic treatment have painted a sobering picture of the ecological disruption that even a single course of antibiotics can cause. A paper published in the journal Nature in 2015 found that a single course of the antibiotic clindamycin reduced the diversity of the gut microbiome significantly and that, two years after the course of treatment, the community had still not fully returned to its pre-antibiotic composition. Certain bacterial species, particularly those that are slow-growing and difficult to culture, appeared to have been permanently displaced. The antibiotic had altered the patient's inner ecosystem in ways that persisted long after the drug had been excreted from the body.

The aftermath of antibiotic treatment also creates windows of vulnerability to opportunistic pathogens. The most dramatic example is Clostridioides difficile, a bacterium that in the presence of a normal, intact gut microbiome is kept in check by competition. When antibiotics devastate the gut community, the ecological competition that normally suppresses C. difficile is removed, allowing it to proliferate to dangerous levels and produce toxins that cause severe diarrhea and colitis. C. difficile infection kills approximately 30,000 Americans per year and is among the most common healthcare-associated infections in the world. It is, in a direct and quantifiable sense, a disease of antibiotic disruption of the microbiome.

The Antibiotic Resistance Crisis

The most globally consequential consequence of antibiotic overuse is not its effect on the microbiome of any individual patient, significant as that is. It is the evolution of antibiotic resistance across entire bacterial populations. When antibiotics are used, they exert intense selective pressure on bacterial communities. The vast majority of bacteria are killed. But a tiny fraction, those that happen to carry genetic variants that confer some degree of resistance to the antibiotic, survive. These survivors reproduce, passing their resistance genes to their offspring. In bacteria, which can reproduce exponentially in hours and can transfer genetic material between unrelated species through a process called horizontal gene transfer, the spread of resistance can be staggeringly rapid.

The result, playing out across the globe over the last several decades, is an accelerating crisis of antibiotic-resistant bacteria. Organisms like methicillin-resistant Staphylococcus aureus, multidrug-resistant Mycobacterium tuberculosis, carbapenem-resistant Enterobacteriaceae, and Neisseria gonorrhoeae resistant to most available antibiotics are now major clinical problems in hospitals and communities worldwide. The World Health Organization has identified antibiotic resistance as one of the greatest threats to global health, food security, and development. Conservative estimates suggest that drug-resistant infections currently kill approximately 700,000 people per year globally. Without significant changes in how we use antibiotics and develop new ones, that number is projected to reach 10 million per year by 2050, exceeding the annual global death toll from cancer.

The Future Beyond Antibiotics

The recognition of the antibiotic crisis has spurred an intense search for alternatives. Bacteriophages, viruses that specifically infect and kill bacteria, represent one of the most promising avenues. Phage therapy was actually developed before antibiotics, in the early 20th century, but was largely abandoned in Western medicine after the introduction of antibiotics due to its perceived complexity and inconsistency. It remained in clinical use in parts of Eastern Europe, particularly in Georgia and Poland, where phage research institutions maintained the expertise. Today, with the antibiotic resistance crisis demanding new solutions, phage therapy is experiencing a renaissance in Western medicine.

Phages offer several advantages over antibiotics. They are exquisitely specific, targeting only the bacteria they are designed to infect and leaving the rest of the microbiome intact. They evolve along with their bacterial hosts, potentially overcoming resistance through the same evolutionary processes that bacteria use to develop it. And they are present in enormous diversity in nature, meaning that for any pathogenic bacterium, there are almost certainly phages that can infect and kill it. Early clinical cases of phage therapy have produced dramatic successes in patients with antibiotic-resistant infections that had not responded to any available drug. The scientific and regulatory pathway to mainstream clinical use remains long and complex, but the promise is real.

PART FIVE

THE FUTURE OF THE MICROBIOME

Rewriting the Code of Human Health

CHAPTER TWELVE

Fecal Microbiota Transplants

The Most Powerful Medicine in the World

There is a medical intervention so effective at treating certain diseases that it produces cure rates exceeding 90 percent in conditions that have defeated antibiotics and conventional medicine. It is backed by rigorous clinical evidence, has been performed safely on tens of thousands of patients worldwide, and works by restoring a disrupted biological system to health rather than simply suppressing symptoms with a drug. And when most people first hear what it actually involves, they react with a visceral combination of disbelief and revulsion.

Fecal microbiota transplantation, the transfer of stool from a healthy donor into the gastrointestinal tract of a sick recipient, is one of the most striking examples in all of medicine of how a treatment that appears, on the surface, to be crude and even disgusting can be, at the level of biology, precisely elegant and profoundly rational. It is also one of the most important tools for understanding what the microbiome actually does, because when you can transplant a community of microbes from one person to another and watch the recipient's health change in predictable and measurable ways, you have moved beyond correlation into something approaching experimental proof.

Ancient Roots, Modern Medicine

The concept of using fecal material as medicine is ancient. A Chinese physician named Ge Hong described the use of human fecal suspensions for the treatment of food poisoning and severe diarrhea in the fourth century CE. This preparation, which he referred to in his writings as yellow soup, was administered orally and was said to produce rapid recovery in patients who had appeared near death. Subsequent centuries of Chinese medical literature contain additional references to similar preparations used for various gastrointestinal conditions. The underlying rationale, that the gut's normal microbial community could be restored through administration of stool from healthy individuals, was intuited long before the concept of the microbiome existed.

Modern fecal microbiota transplantation in Western medicine dates to 1958, when a Colorado surgeon named Ben Eiseman published a case series describing four patients with severe pseudomembranous colitis who were treated with retention enemas of fecal material from healthy donors. All four patients recovered. Eiseman's report attracted little attention for decades, as antibiotics remained effective against the pathogens causing most cases of pseudomembranous colitis. It was only with the emergence of Clostridioides difficile as a major, antibiotic-resistant clinical problem in the 1990s and 2000s that FMT began to receive serious scientific attention.

The C. difficile Revolution

Clostridioides difficile is, as described in the previous chapter, a bacterium that flourishes in the ecological void left by antibiotic destruction of the normal gut microbiome. Its toxins damage the colon lining, causing severe inflammation, watery diarrhea, and sometimes life-threatening colitis. The standard treatment is antibiotics, but here the microbiome-disrupting nature of antibiotics creates a cruel paradox: the antibiotics used to treat C. difficile often further disrupt the already damaged microbiome, creating conditions for relapse. Recurrent C. difficile infection, defined as two or more episodes following treatment, affects up to 35 percent of patients and is associated with dramatically worse outcomes. For patients with recurrent disease, the cure rates with antibiotics drop to around 30 to 40 percent.

Against this dismal backdrop, the results of FMT are extraordinary. The largest randomized controlled trial of FMT for recurrent C. difficile, published in 2013, found that FMT via duodenal infusion cured 81 percent of patients after a single infusion, compared to 31 percent for vancomycin antibiotic therapy. When a second infusion was provided to those who did not initially respond, the overall cure rate rose to 94 percent. These numbers, in a condition where conventional medicine had been failing more than half of patients, were stunning enough to convince the medical establishment that FMT was a legitimate and powerful therapy. The FDA subsequently approved the first standardized FMT product for recurrent C. difficile, transforming what had been an experimental procedure performed differently by different practitioners into a regulated, standardized medicine.

Beyond C. difficile: The Expanding Frontier

The success of FMT in C. difficile has inevitably prompted researchers to investigate whether the same approach might work in other conditions driven by gut microbiome dysbiosis. The pipeline of clinical trials is extraordinary in its breadth, encompassing virtually every condition in which the microbiome is implicated, which, as this book has made clear, is an enormous range of diseases.

For inflammatory bowel disease, including Crohn's disease and ulcerative colitis, FMT has shown promise in multiple randomized trials, though results have been more variable than in C. difficile. The difference likely reflects the greater complexity of inflammatory bowel disease compared to C. difficile infection: it is not simply a matter of restoring a microbial community but of altering a dysregulated immune response in a gut environment that has been chronically inflamed, often for years. Nevertheless, some patients achieve remarkable remissions following FMT, and the identification of clinical and microbiological factors that predict response is an active area of research.

For metabolic conditions including obesity and type 2 diabetes, early FMT trials have produced intriguing if preliminary results. The most famous case is that of patients who developed significant and sustained weight changes after receiving FMT for C. difficile from donors who were either obese or lean. These anecdotal reports, combined with the compelling animal data on microbiome effects on metabolism, have driven formal clinical trials of FMT in obese patients. Results have been modest so far in terms of weight loss, but have revealed interesting insights into the metabolic mechanisms at work.

The Future of FMT: Precision and Pills

The current practice of FMT still involves the transfer of whole, minimally processed stool from a donor to a recipient, a procedure that raises understandable questions about standardization, safety, and scalability. The future of the field is moving rapidly toward more refined and controlled approaches.

Capsule FMT, in which processed and freeze-dried stool from screened donors is encapsulated into odorless, tasteless pills that can be swallowed at home, has already been shown to be as effective as colonoscopic FMT for C. difficile and dramatically more convenient and acceptable to patients. The development of defined bacterial consortia, mixtures of specific bacterial strains selected based on their known functional properties, represents the next step. Rather than transferring the entire, variable microbial ecosystem of a donor, defined consortia would provide a precisely specified set of bacteria known to provide particular therapeutic functions, a much more controlled and reproducible intervention.

The concept of the super donor, an individual whose microbiome is particularly effective at engrafting in recipients and producing therapeutic benefit, is also attracting significant interest. Retrospective analyses of FMT outcomes have consistently found that certain donors produce dramatically better results than others across multiple different recipients, suggesting that some microbiomes are more therapeutically potent than others. Identifying what makes a super donor and developing ways to screen for or create such microbiomes is a major goal of the field.

CHAPTER THIRTEEN

Probiotics, Prebiotics, and Postbiotics

Separating Science from Snake Oil

Walk into any pharmacy or health food store in the industrialized world and you will find an entire aisle dedicated to products promising to improve your gut health, boost your immunity, enhance your mood, support your digestion, and generally transform your biological destiny. These products come in capsules, powders, liquids, gummies, and fermented foods of every description. They carry labels dense with bacterial species names in Latin, with colony-forming unit counts in the billions, with assurances of clinical testing and scientific backing. They are consumed by tens of millions of people worldwide and generate tens of billions of dollars in annual revenue.

How much of this industry is legitimate medicine? How much of it is sophisticated marketing wrapped around essentially empty products? The answer, as with so much in nutrition and health science, is complicated. There are probiotic products and interventions with genuine, replicated, clinically meaningful effects on specific health outcomes. There are also products with little or no evidence of benefit, sold to consumers desperate for solutions to health problems that the products cannot meaningfully address. Learning to tell the difference requires understanding what these terms actually mean and what the science actually shows.

Defining the Terms

A probiotic, according to the formal definition adopted by the Food and Agriculture Organization of the United Nations and the World Health Organization, is a live microorganism that, when administered in adequate amounts, confers a health benefit on the host. This definition contains several critical qualifiers. The organism must be alive at the time of consumption. It must be administered in sufficient quantity to have an effect. And there must be a documented health benefit, established by scientific evidence. Many products that carry the probiotic label fail to meet one or more of these criteria.

A prebiotic is a substrate that is selectively utilized by host microorganisms and confers a health benefit. In practical terms, this means dietary compounds, primarily specific types of fiber, that feed beneficial gut bacteria rather than being absorbed by the host. Examples include inulin, fructooligosaccharides, galactooligosaccharides, and resistant starch. Unlike probiotics, prebiotics do not introduce new organisms into the gut; they provide fuel for the organisms already there. Their effects depend heavily on what is already in the gut microbiome, which is one reason their outcomes are more variable across individuals.

A postbiotic is a more recently defined category referring to the inanimate microorganisms or their metabolic byproducts that confer health benefits on the host. This includes compounds like short-chain fatty acids, exopolysaccharides, cell wall components, and other metabolites produced by bacteria during fermentation. Postbiotics have the advantage over probiotics of not requiring live organisms to be delivered to the gut; the beneficial compounds themselves can be administered directly. This makes them potentially more stable, easier to standardize, and in some cases, more targeted in their effects.

What the Evidence Actually Shows

The scientific evidence on probiotics is both more substantial and more nuanced than either enthusiastic proponents or dismissive skeptics typically acknowledge. For specific conditions and specific strains, the evidence is genuinely compelling. For Clostridioides difficile prevention during and after antibiotic treatment, certain Lactobacillus and Saccharomyces boulardii strains have been shown in multiple randomized controlled trials to reduce the risk of C. difficile diarrhea. For acute infectious diarrhea in children, Lactobacillus rhamnosus GG has been shown in dozens of trials to reduce the duration of diarrhea by approximately one day. For the prevention of necrotizing enterocolitis in premature infants, a life-threatening gastrointestinal emergency, probiotic combinations including Bifidobacterium infantis have been shown to reduce risk significantly. These are real, reproducible clinical effects that justify the use of specific probiotic strains in specific clinical contexts.

The evidence is far weaker for many of the broader claims made about probiotic products in the consumer marketplace. Claims that probiotics boost immunity, improve energy, enhance mood, or treat conditions like irritable bowel syndrome are supported, at best, by weak and inconsistent evidence. The reasons for this weakness are multiple. Most probiotic products contain relatively small numbers of strains, often in concentrations that may not be sufficient to significantly alter the existing microbial community of an adult gut already occupied by hundreds of established species. Many probiotic products are stored and transported in conditions that kill significant fractions of the live bacteria they claim to contain. And the effects of probiotics are highly strain-specific and individual-specific; what works for one person in one condition may have no effect in another person or another condition.

Fermented Foods: The Underrated Champion

While the scientific evidence for many commercial probiotic supplements remains mixed, the evidence for the health benefits of traditional fermented foods has been growing impressively. Fermented foods including yogurt, kefir, sauerkraut, kimchi, miso, tempeh, and kombucha contain live bacterial cultures alongside a rich array of fermentation-derived metabolites, vitamins, and bioactive compounds. They have been consumed by humans across virtually every culture for thousands of years, and their health-promoting effects may reflect this deep co-evolutionary history.

A landmark randomized controlled trial published in Cell in 2021 by researchers at Stanford University compared the effects of a high-fermented-food diet to a high-fiber diet on the gut microbiome and immune markers in healthy adults over ten weeks. The results were striking: the high-fermented-food group showed significant increases in gut microbiome diversity, a marker consistently associated with better health outcomes, and significant decreases in 19 different inflammatory proteins, including markers associated with chronic disease risk. The high-fiber group showed more variable responses, with some individuals showing increased diversity and others showing reduced diversity, suggesting that fiber's effects depend on the existing microbiome composition. The conclusion was clear: increasing consumption of fermented foods produces robust, measurable improvements in gut health that may exceed those achievable with standard probiotic supplements.

Fiber: The Most Underappreciated Medicine

If the science of the microbiome has one clear, consistent, and actionable message for the general public, it is this: eat more fiber. Dietary fiber, the indigestible plant material that feeds gut bacteria and stimulates the production of short-chain fatty acids, is the single most powerful dietary tool available for improving gut microbiome health. The American Gut Project, one of the largest citizen science microbiome studies ever conducted, found that the number of different plant foods consumed per week was the strongest predictor of gut microbiome diversity across all the dietary variables studied. People who ate 30 or more different plant species per week had significantly more diverse microbiomes than those who ate 10 or fewer, and this held true regardless of whether they ate meat or were vegetarian or vegan.

The average American consumes approximately 15 grams of fiber per day, less than half the recommended intake of 25 to 38 grams. The consequences of this fiber deficit for the gut microbiome are significant: bacteria that specialize in fermenting dietary fiber are among the most beneficial organisms in the gut, producing the butyrate and other short-chain fatty acids that maintain gut barrier integrity, reduce inflammation, and support immune function. Without adequate dietary fiber, these species decline, and the microbiome becomes less diverse, less functional, and more susceptible to dysbiosis. Increasing fiber intake is not glamorous advice. It does not involve expensive supplements or exotic interventions. But the evidence that it improves gut health, reduces systemic inflammation, and lowers the risk of colorectal cancer, cardiovascular disease, and metabolic dysfunction is among the most robust in all of nutrition science.

CHAPTER FOURTEEN

The Microbiome Revolution

Personalized Medicine's New Frontier

In a laboratory in Rehovot, Israel, a woman receives what appears to be a fairly ordinary medical evaluation. A nurse takes her blood pressure and temperature. A technician draws blood. She is given a continuous glucose monitor to wear on her arm. She fills out questionnaires about her health history and lifestyle. But she is also asked to do something that no medical evaluation before the twenty-first century would have included: she is given a kit to collect a stool sample, which she will return to the laboratory so that its microbial contents can be analyzed by DNA sequencing.

Over the next few days, this woman's blood glucose is measured continuously as she eats her normal diet. The data is extraordinary in its granularity: every meal, every snack, every cup of coffee is reflected in the continuous glucose trace. And what the data reveals is something that should be obvious in retrospect but that upends the conventional wisdom of nutrition science: her glycemic response to foods is almost entirely personal. Foods that spike her blood glucose dramatically may have almost no effect on the blood glucose of another study participant eating the same food. Foods that barely register for her may cause dramatic spikes in others. The variation is not random. It is patterned. And the patterns correlate strongly with the composition of the gut microbiome.

The End of Universal Nutrition

This study, conducted by Eran Segal, Eran Elinav, and their colleagues at the Weizmann Institute and published in Cell in 2015, enrolled 800 participants and generated one of the most comprehensive datasets ever assembled on the relationship between diet, microbiome, and metabolism. Its central finding challenged one of the foundational assumptions of nutrition science: that food has predictable, generalizable metabolic effects that can be described by universal indices like the glycemic index.

The glycemic index ranks foods by the degree to which they raise blood glucose when consumed in standard amounts. It is the basis of dietary recommendations for millions of people with diabetes and metabolic dysfunction. The Weizmann study found that glycemic responses to identical foods varied enormously between individuals in ways that the glycemic index entirely failed to predict. Some participants had dramatic blood glucose spikes in response to whole grain bread, conventionally considered a low-glycemic food, while experiencing minimal responses to white bread. Others showed the opposite pattern. The same sushi meal, the same banana, the same bowl of rice produced wildly different metabolic effects in different people.

Using machine learning algorithms trained on gut microbiome data, dietary intake, blood parameters, physical activity, and other variables, the researchers were able to develop a personalized predictor of glycemic response that significantly outperformed the glycemic index in predicting how individual participants would respond to specific foods. The predictor was then used to design personalized dietary recommendations for study participants, and in a follow-up trial, these personalized recommendations produced significantly better metabolic outcomes than the standard diet recommended by conventional dietetics.

The message for the future of medicine is clear: the era of universal dietary prescriptions and one-size-fits-all treatment protocols is ending. The microbiome is a critical variable in individual health that conventional medicine has largely ignored, and integrating microbiome data into clinical decision-making will transform the precision and effectiveness of medical and nutritional interventions across a wide range of conditions.

Microbiome Diagnostics: A Window Into Future Disease

The potential to use the microbiome as a diagnostic tool, to read the microbial community and extract clinically actionable information about a person's current and future health, is one of the most exciting frontiers in medical technology. The rationale is straightforward: because the microbiome influences so many aspects of human biology and because its composition is significantly altered in so many disease states, often before symptoms appear, it may contain diagnostic signals that conventional clinical tests miss.

The most advanced clinical application of microbiome diagnostics is in colorectal cancer screening. The current gold standard for colorectal cancer screening is colonoscopy, an invasive, expensive, and patient-averse procedure that requires bowel preparation, sedation, and time off work. Stool-based tests that look for blood or abnormal DNA are less invasive but have significant false positive and false negative rates. Microbiome-based stool tests, which use DNA sequencing to identify the microbial signatures associated with early-stage colorectal cancer and its precursor lesions, have shown impressive sensitivity and specificity in early studies and are moving toward clinical validation.

For conditions beyond colorectal cancer, the diagnostic potential of the microbiome is being explored across a remarkable range of diseases. Researchers have identified microbiome signatures associated with early-stage Parkinson's disease that appear in the gut years before the characteristic neurological symptoms develop. The gut-brain connection in Parkinson's, which often begins with constipation and gut motility problems long before tremor and rigidity appear, has led to the hypothesis that the disease may sometimes originate in the gut and spread to the brain via the vagus nerve, a hypothesis that has accumulated significant supporting evidence in recent years.

Engineered Bacteria: The Next Frontier

Perhaps the most futuristic and yet already rapidly advancing development in microbiome medicine is the engineering of bacteria to perform therapeutic functions within the body. Synthetic biology, the discipline that combines molecular biology and engineering to design and build biological systems with novel functions, has made possible the creation of bacteria that can sense disease states in the gut, produce therapeutic proteins in response, and deliver those proteins exactly where they are needed.

Researchers at MIT, Harvard, and several biotechnology companies have engineered strains of bacteria that can detect the biomarkers of intestinal inflammation and respond by producing anti-inflammatory compounds. Other engineered bacteria have been programmed to sense the presence of cancer-associated compounds in the gut and produce signals detectable in the urine, offering the possibility of a non-invasive cancer screening test that could be performed simply by drinking a probiotic and then collecting a urine sample. Still others have been designed to produce missing metabolites in patients with metabolic genetic disorders, offering a potential microbiome-based approach to diseases that currently require complex dietary management or enzyme replacement therapy.

These therapeutic bacteria, sometimes called living medicines, represent a fundamentally new paradigm in pharmacology. Unlike conventional drugs, which act on the body in predictable but often blunt ways, living medicines can sense the local environment, integrate multiple signals, and respond with precisely calibrated outputs. They can be designed to persist in the gut for specific durations, to respond only under specific conditions, and to be eliminated when treatment is complete. The technical and regulatory challenges of developing these therapies are formidable, but the potential rewards, treatments for conditions that currently have no adequate therapeutic options, are compelling enough to drive substantial investment and scientific effort.

Equity and Access in the Microbiome Age

The revolution in microbiome medicine raises a question that must be asked alongside the celebration of scientific progress: who will benefit? The history of medical innovation is, unfortunately, also a history of inequitable access. Expensive new therapies tend to be available first to those with wealth and good health insurance. The personalized medicine revolution, which promises treatments tailored to the individual biology of each patient, risks becoming a revolution available only to individuals whose biology and financial resources make them attractive to healthcare systems motivated by profit.

The microbiome itself is already a measure of social inequality. Studies consistently find that people living in poverty, in food-insecure households, in neighborhoods with limited access to fresh produce and green space, and with elevated exposure to environmental toxins have less diverse gut microbiomes than their more affluent counterparts. The factors that impoverish the microbiome, ultraprocessed food, antibiotic overuse, environmental pollution, chronic stress, inadequate sleep, limited access to nature, are disproportionately concentrated among people with fewer economic resources. The diseases that result from a degraded microbiome, obesity, metabolic syndrome, inflammatory bowel disease, mental health disorders, are more prevalent among the economically disadvantaged. A microbiome medicine that only treats those who can afford expensive sequencing, personalized therapies, and specialty probiotic formulations while ignoring the social determinants of microbial health will widen rather than narrow the health inequalities that already exist.

CHAPTER FIFTEEN

Rewilding Your Inner Universe

How to Protect and Restore Your Microbiome

This final chapter is a conversation. Not a lecture, not a list of commandments, not a rigid protocol to be followed with militaristic precision. The science of the microbiome, as you now know deeply and in full complexity, does not support rigid universalism. The microbiome is individual. Its needs are individual. And the interventions that will nourish it are individual too. What I offer here is not a prescription but a framework: a way of thinking about the choices you make every day, from what you eat to how you sleep to where you spend your time, that takes seriously the biological reality of the living ecosystem you carry within you.

The concept of rewilding comes from conservation ecology. It refers to the restoration of degraded ecosystems by reintroducing the species, processes, and ecological relationships that once characterized them in their healthy state. Rewilding a landscape does not mean turning back the clock to some imaginary pristine past. It means recognizing what has been lost, understanding why it was lost, and making deliberate choices to restore the ecological richness that makes an ecosystem resilient, productive, and capable of sustaining life. The same principle applies, with remarkable directness, to the human gut microbiome.

Food as Ecological Management

The most powerful tool you have for shaping your microbiome is the most obvious one: food. What you eat, in a very literal sense, determines who gets to live in your gut. Feed fiber-fermenting bacteria with dietary fiber and they flourish, producing the short-chain fatty acids that protect your gut lining, reduce inflammation, and support your immune and metabolic health. Deprive them of fiber and they decline, leaving ecological space for less beneficial organisms. The relationship between diet and microbiome is so rapid and so responsive that measurable changes in microbial community composition can be detected within 24 to 48 hours of a major dietary shift. Your microbiome is not a static entity fixed by your genetics. It is a dynamic ecosystem that responds to your daily choices.

The most consistent finding in dietary microbiome research, as noted in the chapter on probiotics and prebiotics, is that plant diversity drives microbiome diversity. The American Gut Project's finding that eating 30 or more different plant foods per week is the strongest predictor of gut microbiome diversity is one of the most actionable results in the field. Thirty different plant foods per week sounds like a lot, but it is more achievable than it appears: herbs and spices count, as does every different type of vegetable, fruit, legume, nut, seed, and whole grain that you eat. A salad with five different vegetables, topped with three different nuts and seeds, seasoned with two different herbs, already contains ten plant types. Adding variety to an existing healthy diet can reach 30 plants per week without any radical transformation of eating habits.

Fermented foods deserve particular emphasis, given the strength of the evidence reviewed in earlier chapters. Incorporating at least one serving of traditionally fermented food daily, whether yogurt, kefir, sauerkraut, kimchi, miso, tempeh, or another fermented product appropriate to your cultural background and taste preferences, provides both live bacterial cultures and a rich array of fermentation-derived metabolites that support gut health. The most important criterion is that the product contains live cultures: heat-treated or heavily preserved fermented foods may have lost their beneficial bacteria and offer primarily the flavor rather than the microbial benefit of the fermentation process.

Ultra-processed foods, defined as industrial formulations containing food-derived ingredients combined with additives, flavors, and preservatives not used in domestic cooking, are among the most damaging influences on the gut microbiome. They are typically low in fiber and high in refined carbohydrates, sugar, and unhealthy fats, providing poor nutritional support for beneficial bacteria. Many contain emulsifiers like carboxymethylcellulose and polysorbate 80 that have been shown in animal studies to disrupt the mucus layer protecting the gut lining and to alter microbiome composition in ways that increase intestinal permeability and inflammation. Artificial sweeteners, as discussed earlier, alter the microbiome in ways associated with metabolic dysfunction. Reducing or eliminating ultra-processed foods from the diet is one of the most impactful steps most people in industrialized nations can take for their gut health.

Movement, Nature, and the Outdoor Microbiome

Exercise has profound and well-documented effects on the gut microbiome. Multiple studies in both humans and animal models have shown that regular aerobic exercise increases the abundance of butyrate-producing bacteria, increases overall microbial diversity, and alters the microbiome in ways that reduce systemic inflammation. A study published in Gut in 2018 found that professional rugby players had significantly more diverse gut microbiomes and higher abundance of several beneficial bacterial species than sedentary controls, even after controlling for dietary differences. The effect of exercise on the microbiome appears to be independent of diet: active individuals have healthier microbiomes than sedentary individuals even when they eat similar foods.

The mechanisms by which exercise influences the microbiome are still being worked out, but several pathways have been identified. Exercise increases gut motility, the rate at which contents move through the digestive system, which affects which bacteria have time to establish themselves in different regions of the gut. It affects the gut's immune environment, changing the types and activation states of immune cells that interact with gut bacteria. It influences the production of bile acids, which serve as antimicrobial agents in the gut and as signaling molecules that bacterial populations respond to. And it changes the availability of substrates for fermentation, as muscle metabolism during exercise produces lactic acid and other compounds that reach the gut and influence its microbial ecology.

Beyond structured exercise, time spent in natural environments has specific and well-documented benefits for the microbiome. Green spaces, forests, gardens, and farms are rich in environmental microbes that diversify the human skin and gut microbiome when we spend time in them. The Finnish immunologist Ilkka Hanski found that children who lived near biodiverse natural environments had more diverse skin microbiomes and lower rates of allergic sensitization than children who lived near species-poor, degraded landscapes. A study of urban children found that those whose school yards were converted from conventional asphalt surfaces to biodiverse garden environments showed changes in skin microbiome composition and immune markers within weeks. The environment you inhabit shapes the microbes you encounter, and the microbes you encounter shape your immune system.

Sleep as a Microbial Prescription

Sleep hygiene, the set of practices that promote regular, restorative sleep, is most commonly discussed in the context of cognitive performance and mental health. The microbiome science adds another compelling reason to take sleep seriously. As discussed in the chapter on stress and sleep, the gut microbiome follows its own circadian rhythms synchronized with the host's sleep-wake cycle. Irregular sleep patterns disrupt these microbial rhythms, altering the microbiome's composition and reducing its functional effectiveness. Conversely, regular, adequate sleep of seven to nine hours per night supports microbial circadian synchronization and promotes the diversity and stability of the gut microbial community.

Practical sleep hygiene recommendations from microbiome science align closely with those from sleep medicine more generally: maintaining a consistent sleep schedule seven days a week, creating a dark, cool, and quiet sleep environment, avoiding blue light exposure from screens in the hour before bedtime, and moderating alcohol and caffeine consumption that interferes with sleep architecture. These recommendations are not new. What the microbiome science adds is an additional biological rationale, an understanding that poor sleep harms not just the brain and the cardiovascular system but the entire inner ecosystem whose health determines so much else about your wellbeing.

Managing Antibiotics and Medications Wisely

Understanding the microbiome-disrupting effects of antibiotics does not mean avoiding them when they are genuinely needed. Antibiotics save lives. When a bacterial infection is serious and requires antibiotic treatment, the decision to use antibiotics is straightforward. What microbiome science encourages is more careful deliberation about antibiotic use in situations where their necessity is less clear: mild respiratory infections that are most commonly viral and do not benefit from antibiotics, cases where watchful waiting is clinically appropriate, contexts where the choice between a narrow-spectrum antibiotic with limited microbial disruption and a broad-spectrum one with extensive collateral damage can be made thoughtfully with a physician's guidance.

When antibiotics are necessary, supporting the microbiome during and after treatment can help accelerate recovery of the microbial community. Taking a probiotic containing Lactobacillus rhamnosus GG or Saccharomyces boulardii alongside antibiotic treatment has been shown in multiple trials to reduce the risk of antibiotic-associated diarrhea, though the probiotic should ideally be taken several hours away from the antibiotic to avoid being killed before it can establish in the gut. Prioritizing fiber-rich and fermented foods during and after antibiotic treatment provides the substrates and organisms that the recovering microbiome needs to rebuild. And avoiding additional unnecessary microbiome stressors, ultra-processed foods, excess alcohol, high stress, and inadequate sleep, in the period following antibiotic treatment gives the ecological recovery process the best possible conditions.

The Long View: Your Microbiome Is Your Legacy

The microbiome is not just a personal health issue. It is a generational one. The microbial communities that parents carry influence the microbiomes of their children from the very beginning, through the birth process, through breastfeeding, through the shared household environment, through the diet and lifestyle choices that shape the family's microbial ecology. The choices you make about your own microbiome have implications not just for your health but for the health of the children you may have or already have, and through them, for the health of their children in turn.

This is both a responsibility and an opportunity. The degradation of the human microbiome in industrialized nations over the last century, through the combined forces of antibiotic overuse, dietary impoverishment, loss of contact with natural environments, and rising chronic stress, is real and consequential. It cannot be reversed overnight. But it can be addressed, deliberately and with the guidance of increasingly sophisticated science, one generation at a time. Rewilding the human microbiome is not a personal project. It is a civilizational one, and it begins with each individual choice to feed the invisible universe within.

You are, as I wrote in the preface, not one person. You are an ecosystem. An ancient, complex, remarkable ecosystem, shaped by billions of years of evolutionary partnership between human and microbial life. The trillions of organisms that share your body are not parasites or passengers. They are partners. They have made you what you are, and what you make of yourself will shape what they become. That is not a burden. It is, if you choose to see it clearly, one of the most profound and beautiful facts about being alive.

EPILOGUE

What It Means to Be Human

We began this book with microbes. We end with a question. What does it mean to be human in a world where the boundaries of the self are not where we thought they were?

The traditional view of the individual, the self-contained person with a clear biological boundary at the skin, has been one of the organizing fictions of modern medicine, modern philosophy, and modern culture. We speak of the individual as the fundamental unit of health, of responsibility, of identity. We treat the human body as a thing that can be understood and repaired in isolation from the microbial world that inhabits it. We make choices, take medications, eat foods, and structure our environments as though the only biology that matters is the 20,000 genes in our human genome.

The microbiome science dismantles this fiction gently but completely. We are not individuals. We are ecosystems. The organisms that share our bodies are not incidental accompaniments to human life. They are integral to it, evolutionarily co-authored with it, functionally inseparable from it. Our immune systems were built with their help. Our moods are shaped by their chemistry. Our metabolism runs partly on their metabolites. The very architecture of our development, from the moment of birth through the education of our immune systems in childhood, bears the mark of their influence.

Does this diminish us? Only if you believe that being complex is somehow less dignified than being simple, that being interconnected is somehow less noble than being self-sufficient. I believe the opposite. The discovery that we are ecosystems, that our existence is a cooperative achievement of billions of years of co-evolution between human and microbial life, is not a reduction of human significance but an expansion of it. We are more interesting, more intricate, more deeply woven into the fabric of the living world than we knew. Our health is not just our own. Our biology is not just our own. We are connected, in the most intimate biological sense, to the ancient microbial world from which all complex life ultimately descends.

That connection is worth knowing. It is worth protecting. And it is worth passing on, to our children, to their children, and to the generations who will inherit the world we leave behind. The microbes were here first. With luck and wisdom, they will still be flourishing when the last of our distant descendants draws their last breath on this remarkable, microbially sculpted planet.

Mohammad Umar

2026

APPENDICES

Supporting Material and Resources

Appendix A: Glossary of Key Scientific Terms

Archaea: A domain of single-celled microorganisms distinct from bacteria and eukaryotes, often found in extreme environments. Some species inhabit the human gut where they play roles in metabolism, particularly in the breakdown of hydrogen gas produced during fermentation.

Bacteriome: The bacterial subset of the microbiome, referring specifically to the bacterial communities that inhabit a given environment such as the gut, skin, or mouth.

Bacteroidetes: One of the two dominant bacterial phyla in the adult human gut microbiome, alongside Firmicutes. Species within this phylum are important fiber digesters and are generally associated with lean body composition.

Bifidobacterium: A genus of gram-positive bacteria particularly important in the infant gut, where they dominate the microbial community in breastfed babies. They are among the most commonly used organisms in commercial probiotic products.

Butyrate: A short-chain fatty acid produced by the bacterial fermentation of dietary fiber in the colon. It is the primary energy source for colonocytes, has powerful anti-inflammatory properties, and plays a critical role in maintaining gut barrier integrity and colon health.

Colonocyte: The epithelial cell that lines the colon. Colonocytes are critically dependent on butyrate produced by gut bacteria for their energy needs and survival.

Dysbiosis: An imbalance or disruption in the composition of a microbial community, particularly the gut microbiome, that is associated with disease states. Dysbiosis can involve loss of diversity, depletion of beneficial species, or overgrowth of harmful ones.

Enteric nervous system: The extensive network of neurons embedded in the walls of the gastrointestinal tract, sometimes called the second brain. It contains over 500 million neurons and can function independently of the central nervous system.

Fecal microbiota transplant (FMT): A medical procedure in which stool from a healthy screened donor is transferred into the gastrointestinal tract of a patient, with the goal of restoring a healthy microbial community. Currently FDA-approved for recurrent Clostridioides difficile infection.

Firmicutes: One of the two dominant bacterial phyla in the adult human gut microbiome. This phylum contains many important fiber-fermenting species, including the butyrate producers that are critical for gut health. The Firmicutes to Bacteroidetes ratio has been associated with metabolic health.

GABA (gamma-aminobutyric acid): The primary inhibitory neurotransmitter in the brain, which reduces neuronal excitability. Some gut bacteria produce GABA, which influences gut-brain signaling through the vagus nerve.

Gut-brain axis: The bidirectional communication network connecting the gut and the brain, involving neural pathways (primarily the vagus nerve), hormonal signals, immune signals, and microbial metabolites.

Human Microbiome Project: A landmark NIH-funded research initiative launched in 2007 to characterize the microbial communities of 300 healthy adults across 18 different body sites, providing the foundational reference dataset for human microbiome science.

Leaky gut: A colloquial term for increased intestinal permeability, a condition in which the tight junctions between gut epithelial cells become loosened, allowing bacterial products to enter the bloodstream and trigger systemic inflammation.

Metagenome: The collective genetic material of all organisms in a given microbial community. The human metagenome, encompassing all the genes of all microorganisms in the body, contains approximately 150 times more genes than the human genome itself.

Microbiome: The collection of all microorganisms, including bacteria, viruses, fungi, archaea, and other microorganisms, along with their collective genetic material, that inhabit a particular environment such as the human body.

Microbiota: The community of microorganisms themselves, as distinct from the microbiome which includes their genetic material.

Old friends hypothesis: A refinement of the hygiene hypothesis, proposing that the human immune system requires exposure not simply to infections but to the specific organisms it has co-evolved with over millions of years, including environmental bacteria, helminths, and other microbes, in order to develop and function normally.

Postbiotic: Inanimate microorganisms or their metabolic byproducts that confer health benefits on the host, including short-chain fatty acids, exopolysaccharides, and other fermentation products.

Prebiotic: A substrate that is selectively utilized by host microorganisms and confers a health benefit. Dietary fibers such as inulin, fructooligosaccharides, and resistant starch are the primary prebiotics.

Probiotic: A live microorganism that, when administered in adequate amounts, confers a health benefit on the host. Common probiotic organisms include Lactobacillus and Bifidobacterium species.

Psychobiome: The subset of the gut microbiome with particular relevance to brain function, mental health, and behavior, including organisms that produce or regulate neurotransmitters and influence the gut-brain axis.

Short-chain fatty acids (SCFAs): Metabolites produced by the bacterial fermentation of dietary fiber in the colon, primarily butyrate, acetate, and propionate. They are essential for gut health, immune regulation, and metabolic function.

Vagus nerve: The tenth cranial nerve, running from the brainstem through the neck, chest, and abdomen. It is the primary neural pathway of the gut-brain axis, carrying signals in both directions, with approximately 80 percent of traffic flowing from the gut to the brain.

Appendix B: Key Scientists and Institutions

The science described in this book stands on the shoulders of hundreds of researchers around the world. The following are among the most important contributors to the fields covered in these pages. Jeffrey Gordon at Washington University in St. Louis, whose germ-free mouse experiments and human microbiome studies have been foundational to understanding the microbiome's role in metabolism and obesity. John Cryan and Ted Dinan at University College Cork, pioneers of the gut-brain axis and psychobiome fields. Eran Segal and Eran Elinav at the Weizmann Institute of Science in Israel, whose personalized nutrition research has transformed understanding of individual metabolic variation. Sarkis Mazmanian at Caltech, whose work on the immune-modulating effects of gut bacteria has been transformative. Rob Knight at the University of California San Diego, who co-founded the American Gut Project and has driven the democratization of microbiome science. Graham Rook at University College London, developer of the old friend's hypothesis. Peter Turnbaugh, whose work on the microbiome and metabolism at the University of California San Francisco has been enormously influential. Maria Gloria Dominguez-Bello, whose research on birth mode and the infant microbiome has changed our understanding of neonatal microbial colonization.

Appendix C: A 30-Day Microbiome Reset Framework

This framework is not a rigid protocol but a progressive set of changes that, implemented gradually over 30 days, can meaningfully improve gut microbiome health. Consult with a healthcare provider before making significant dietary changes, particularly if you have existing health conditions.

During the first week, the focus is on assessment and foundation. Begin by keeping a food diary to understand your current fiber intake and plant food diversity. Introduce one serving of fermented food daily, whether yogurt, kefir, sauerkraut, or kimchi. Begin replacing one refined carbohydrate serving per day with a whole grain or legume alternative.

During the second week, expand plant diversity. Set a goal of including 15 different plant foods this week. Experiment with one unfamiliar vegetable, fruit, or grain. Add a handful of mixed nuts and seeds to your daily eating.

During the third week, address lifestyle factors. Establish a consistent sleep schedule and protect sleep duration. Incorporate 30 minutes of moderate aerobic exercise at least five days this week. Spend time in a natural green environment at least twice.

During the fourth week, consolidate and expand. Aim for 25 to 30 different plant foods this week. Evaluate how you feel and note any changes in digestion, energy, mood, or sleep quality. Consider the sustainability of these changes and identify which are most feasible to maintain long-term.

Appendix D: Further Reading

Ed Yong's book I Contain Multitudes remains the most comprehensive and beautifully written popular science account of the microbiome and deserves to be read by anyone who wants to go deeper into the science covered in these pages. Martin Blaser's Missing Microbes makes a compelling case for the microbiome-depleting consequences of antibiotic overuse and is essential reading on that subject. The Good Gut by Justin and Erica Sonnenburg of Stanford University offers practical guidance grounded in solid science. For those with a more technical appetite, the journals Cell, Nature, Science, and Cell Host and Microbe regularly publish the foundational research in this field. The website of the NIH Human Microbiome Project provides access to publicly available data and research summaries. The Sonnenburg lab at Stanford and the Gordon lab at Washington University in St. Louis maintain research websites with accessible summaries of their ongoing work.